ENCYCLOPAEDIA OF PLANT PROTECTION

ENCYCLOPAEDIA OF
PLANT PROTECTION
Vol. 3

B.A. Mannan

ANMOL PUBLICATIONS PVT. LTD.
NEW DELHI - 110 002 (INDIA)

ANMOL PUBLICATIONS PVT. LTD.
H.O.: 4374/4B, Ansari Road, Darya Ganj,
New Delhi-110 002 (India)
Ph.: 23278000, 23261597
B.O.: No. 1015, Ist Main Road, BSK IIIrd Stage
IIIrd Phase, IIIrd Block,
Bangalore - 560 085 (India)
Visit us at: www.anmolpublications.com

Encyclopaedia of Plant Protection

First Published, 2008

ISBN 978-81-261-3538-7 (Set)

PRINTED IN INDIA

Printed at Mehra Offset Press, Delhi.

Contents

PREFACE

Plant lovers, food producers and agricultural planners, all voice together to protect the plants from insects and diseases. Damages to the plants by insects are easily felt, but not realised fully. Every year, hundreds of people die of hunger. We can provide them with enough, food, if food crops are kept safe from the damages.

Plants, like human beings, are prone to different diseases. A diseased plant cannot yield good produces. Hence, plant protection has attained greater significance in today's world. Agricultural crops have been prone to damage, because of insects, rodents, diseases, weeds and nematodes, etc. since time immemorial.

Plant Protection continues to play a significant role in achieving targets of crop production. Major thrust areas of plant protection are promotion of integrated pest management, ensuring availability of safe and quality pesticides for sustaining crop production, despite the ravages of pests and diseases by streamlining the quarantine measures for accelerating the introduction of new high-yielding crop varieties, besides eliminating the chances of entry of exotic pests.

It is essential to have the basic knowledge about the plant pathogens, different kinds of diseases, caused by them to various crops and of course, the methods of controlling diseases to avoid enormous losses caused by the diseases. As plant protection is indispensable, job opportunities are always there for people, having basic knowledge of pests, crops and diseases.

This book is a serious endeavour in that direction. This *Encyclopaedia of Plant Protection,* in four volumes takes the stock of present state of art in plant protection. It also focuses on isolation, structure, characterisation and physiochemical and physiological properties of plants, synthesis, functions and mechanisms of action of plant protease inhibitors. It also deals with their inhibitory effects on insects and pathogens and with the resultant application of genetic engineering to plants for their protection.

This book caters to the need of knowledge hungry biologists, agriculturists, policy planners and specialists in the field of plant protection and of course the students, teachers, researchers and beyond that, all those concerned with agricultural production and its development.

Editor

1

MANAGING PESTS

The unceasing struggle between man and his insect enemies started even before the dawn of civilization. In spite of the numerous advances made by man in evolving newer and deadlier weapons to fight the war against insects, he has not succeeded in eradicating even one of the thousands of serious pests which damage his food and other agricultural products, destroy his possessions and even attack himself and injure his domestic animals.

There generally exists an uneasy truce between the insect pests and man, and this is termed as 'balance in nature. This balance is the result of two opposing phenomena, the 'biotic potential, i.e. the tremendous capacity of insects to reproduce and multiply and the 'environmental resistance' which keeps their numbers under check. The environmental resistance results in the death of adults before oviposition, in the mortality of the eggs, larvae or pupae of the insects because of desiccation, starvation, parasites, predators, diseases and other adverse environmental conditions. Even a slight slackening of any of the processes of 'environmental resistance' results in a population explosion of an insect species and the consequent epidemic.

The change in 'environmental resistance' may take place owing to a number of causes, either natural or operated by different agencies. Man is, perhaps, the single most important agent, who has, from time to time, disturbed the 'balance in nature', and this has caused numerous pest problems and pest epidemics. The nature, too, plays an important role in causing pest epidemics. Favourable conditions which reduce the natural mortality and bring about the rapid development of the insect coupled with the conditions unfavourable to the natural enemies of that insect, often result in a rapid increase of its population leading to a sudden pest out-break.

Different Lookouts

Once it is established that an insect is causing *economic losses,* it becomes necessary to control it.

The first principle underlying the control of an insect is its correct identification. When it is correctly identified, we can refer to the available information on the biology and habits of that insect and determine its most vulnerable stage, the appropriate time and the most suitable method or methods to control it. The knowledge and understanding of the ecological factors, both biotic and abiotic, affecting the population of an insect pest is necessary for planning the proper strategy for controlling it. A lot of attention is being paid to this aspect now and elaborate procedures and 'models' have been evolved to pinpoint the 'key factors', dominantly affecting the development and multiplication of a particular pest.

The choice of the proper method or methods of control becomes easier, when the above-mentioned information becomes available in respect of a particular pest. Some of the important methods for the control of insect pests are discussed below.

Physical and Mechanical Control Method: This is one of the oldest methods, and has been in use since time immemorial. It includes measures, such as the collection of egg masses and other inactive stages, the removal and destruction of infested parts or whole plants, the beating of drums or tins and trenching. This method can prove to be effective during the initial stage of the pest incidence and when practised as a concerted effort by a large number of farmers in a particular area.

Cultural Control Method: This is a preventive method which is inexpensive and may prove effective and efficient, if employed after acquiring a thorough knowledge of the life-history and habits of a pest. Deep ploughing after harvesting a crop (to expose the hiding or resting insects), the removing and destroying of the stubble and other trash, adjusting the time of sowing (to avoid the peak-incidence period), clean cultivation, the removal of alternative wild hosts, catch crops and suitable rotations are some of the important measures included under this method of control.

Resistance of Host Plant: Host-plant resistance has generally been considered one of the components of the cultural method, but because of its importance, it deserves an independent status as a major method of protecting crops against insect damage. The method involves the utilization of the inherent property of certain strains or varieties of crops, of being less infested or less damaged than other strains or varieties of the same crop. Enormous economic benefits have been achieved by using this method in the case of crops, e.g. wheat, maize, cotton and alfalfa. Very little attention has so far been paid in India to the exploitation of this method. However, under the various coordinated projects on the improvement of any crop, special attention is being given to the evolvement of varieties resistant to major insect pests. It is realized that the success of this method requires the close cooperation of entomologists, geneticists and plant-breeders.

Biological Control Method: Practically every crop pest has its natural enemies in the form of parasites, predators and disease-causing organisms. The biological control involves a large-scale multiplication and liberation of such agents, or creating conditions under which the naturally occurring agents can act effectively. This type of control cannot be undertaken by individual farmers and has necessarily to be carried out by specialized agencies.

Some very outstanding successes have been achieved by using this method, but the method suffered a set-back owing to the large-scale and indiscriminate use of insecticides. The approach at present is to evolve methods by which, the biological and chemical methods can be integrated, so that the harmful effects of insecticides do not interfere with the activities of the natural enemies.

Legislative or Regulatory Control Method: Legislative or regulatory method is a specific method mainly employed to prevent the introduction of pests from other countries or to prevent the spread of a pest from one area to another. The method is operated through specific regulations known as plant-quarantine laws. In India, some states have special pest acts, by which it becomes obligatory on the part of cultivators and governmental authorities to take appropriate steps to control a particular pest when it appears in an epidemic form.

Chemical Control Method: After the discovery of the insecticidal properties of DDT in 1939 by Paul Muller, the chemical control of insects has become most popular. As a matter of fact, the method is so popular that most of the cultivators and extension workers speak only about insecticides, whenever the question of insect control is raised. The main reason for its popularity is the spectacular and immediate results obtained by the use of such chemicals.

Hundreds of insecticides are now available to control different insect pests. These are used mainly as dusts, sprays or granules on the crops; dusts and granules can also be incorporated into the soil for the control of soil-inhabiting insects. In recent years, the soil application of some of the systemic insecticides has proved to be effective in controlling insects attacking crops in the early stages of their growth.

Some of the serious limitations of the insecticides have been highlighted in recent years. Out of these, the problems of insecticide residues on crops and other products used as food and fodder and on the pollution of environment have been agitating the minds of people in the developed and developing countries of the world. This realization has resulted in certain extremist quarters demanding the banning of the use of most of the insecticides. However, the sensible approach suggested and generally accepted is for a judicious and restricted use of insecticides. It is strongly felt that in the near future, there is no possibility of replacing the chemical method of controlling insects entirely by any other method.

Various Procedures

In addition to the major methods of controlling insects described above, there are a number of other methods, which have either been employed successfully for controlling certain specific pests, or are still in the experimental and developmental stages. Increased attention is being paid now to the incorporation of such methods in the programmes of insect control. Some of these methods are: the use of attractants, repellents and anti-feeding compounds, radiations for sterilization (the well-known example of eradicating the screw-worm by using the sterile-male technique), chemosterilants, juvenile and sex hormones. These methods can in future replace or supplement the use of insecticides.

Insect Control

The guiding principles for selecting an appropriate method or methods of control should be that the method chosen must be economical, free from creating any other problem, immediately or in future, it should not harm the natural enemies of the pest and should be easy to operate and be readily available to an ordinary cultivator.

The choice of the proper method (or methods) of control becomes easier when the biology and habits of the pest are known and its most vulnerable stage has been determined. However so far the general tendency among those who are responsible for carrying out the control operations is to employ an easy method, giving quick results and disregarding other unfavourable consequences which may follow the completion of the operations.

Keeping in view the guiding principles in the selection of pest-control methods, it is now generally realized that any single approach to the problem of insect control is not feasible. Each method has its advantages as well as its disadvantages and also has limitations under a particular set of conditions. The current thinking all over the world is to suitably synthesize the use of as many methods as possible to control the insect pests effectively, economically and without any adverse after-effects. This approach is now popularly known as 'Integrated Pest Control'.

Despite a number of serious limitations from which insecticides suffer, it is felt that the use of these chemicals in the control of insect pests will continue for a long time to come and in the strategy for 'Integrated Pest Control', this method will occupy an important place as a curative method for reducing losses because of pests.

However, the emphasis is to be laid on the judicious and proper use of insecticides, so that this method may not impair the efficacy of other methods and create other problems. Keeping this point in view, some more information on insecticides and their proper use is given in the following pages.

Various Insecticides

The concept of using chemicals for controlling insect pests is not new; however, its importance and popularity increased only after the discovery of the insecticidal properties of DDT. Originally, these chemicals were classified on the basis of their mode of entry in the bodies of insects, viz. stomach poisons, contact poisons and fumigants. However, this classification has become outdated, because most of the synthetic organic insecticides act both as stomach and contact poisons and some have also the added fumigant action. At present, the generally accepted classification is based on the chemical nature of the insecticides.

Division of insecticides (based on their chemical nature)

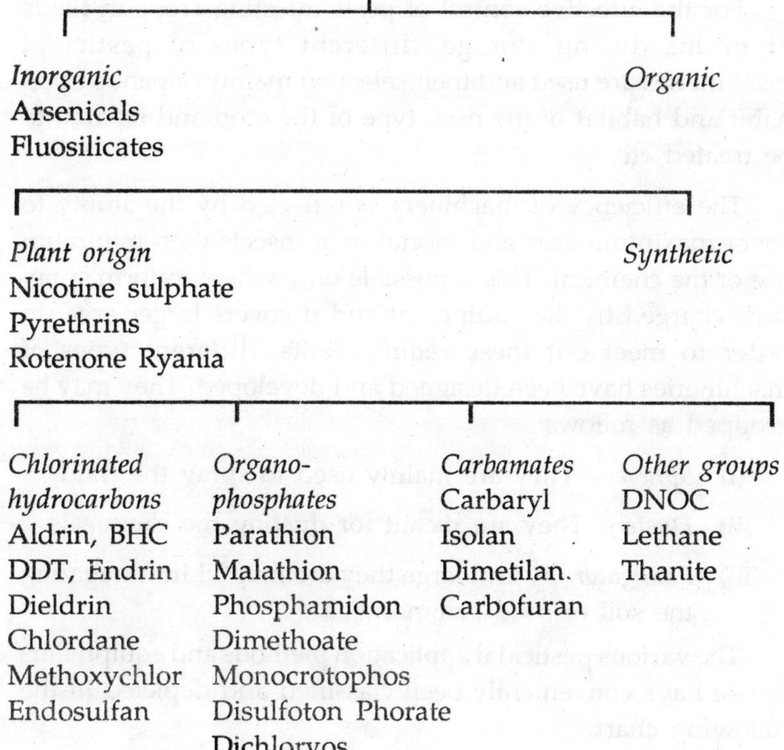

In Case of Skin Contamination

It is just possible that during the handling of pesticides some parts of the body may come in the contact of poison. Under such circumstance following treatments may be followed:

1. The contaminated part may be washed with soap immediately.
2. The related part of the body should be washed for sometime with running water.
3. If necessary, Burnol may be used to avoid irritation.

THE MACHINARY

Insecticidal Machinery

For the effective control of pests infesting crops in fields or grains during storage, different types of pesticidal machineries are used and their selection mainly depends upon habit and habitat of the pest, type of the crop and its area to be treated etc.

The efficiency of machinery is reflected by the ability to cover maximum area and mortality of insects with minimum use of the chemical. This is possible only when uniform spray is discharged by the equipment and it covers larger area. In order to meet out these requirements, different types of machineries have been designed and developed. They may be grouped as follows:

(i) *Sprayers:* They are mainly used to spray the fluid.

(ii) *Dusters:* They are meant for dusting the chemicals.

(iii) *Fumigator:* By and large they are helpful in fumigating the soil or food commodities.

The various pesticidal application methods and equipments in use have conveniently been classified and depicted in the following chart.

Insecticides Toxicity: Practically all insecticides are toxic to man and other animals; however, they differ in their degree of toxicity. Thus the hazards to man and his domestic animals depend on the values and nature of toxicity (oral or dermal).

The United States Environmental Protection Agency, Washington, has recognized three categories, viz. danger-poison, warning and caution on the basis of these values. Available data on some of the common insecticides are presented in Table.

Table. Acute Oral LD50 Values of Some Common Insecticides for White Rats

Insecticides	*Oral LD_{50} (mg/kg)*		*Remarks**
	Males	*Females*	
(A) Chlorinated hydrocarbons			
Aldrin	39	60	Toxic
Benzene hexach-loride (BHC)	1,250	—	Less toxic, sax not mentioned
Chlordane	335	430	Toxic
DDT	113	118	Less toxics
Dieldrin	46	46	Toxic
Endosulfan	43	18	Toxic
Endrin	17.8	7.5	Highly toxic
Heptachlor	100	162	Toxic
Lindane (gamma BHC)	88	91	Toxic
Toxaphene	90	80	Toxic

Contd...

Insecticides	*Oral LD_{50} (mg/kg)* *Males*	*Females*	*Remarks**
(B) Organo phosphates			
Aldicarb	0.93		Highly toxic, sex not mentioned
Diaxinon	108	76	Toxic
Dichlorvos	80	56	Toxic
Dimethoate	215	—	Toxic
Disulfoton	6.8	2.3	Highly toxic
Fenitrothion	740	—	Less toxic, sex not mentioned
Malathion	1,375	1,000	Less toxic
Methyl parathion	14	24	Highly toxic
Mevinphos	6.1	3.7	Highly toxic
Monocrotophos	17.5	20	Highly toxic
Parathion	13	3.6	Highly toxic
Phorate	2.3	1.1	Highly toxic
Phosphamidon	23.5	23.5	Highly toxic
(C) Carbamates			
Carbofuran	11	—	Highly toxic, sex not mentioned
Carbaryl	850	500	Less toxic
Zectran	37	25	Toxic
(D) Plant origin			
Nicotine sulphate	—	83	Highly toxic

Contd...

Insecticides	*Oral LD_{50} (mg/kg)* *Males*	*Females*	*Remarks**
Pyrethrins	1,500	—	Less toxic, sex not mentioned
Pyrethrum	1,870	820	Less toxic
Rotenone	50-75		Less toxic, sex not mentioned
Ryania	1,200	—	Less toxic, sex not mentioned
(E) ***Inorganic***			
Calcium arsenate	—	298	Less toxic
Lead arsenate	—	1,050	Less toxic
Paris green	—	100	Less toxic

Use of Insecticides and Precautionary Measures: The improper, excessive and careless use of insecticides may prove to be injurious to man and his domesticated animals. Proper safeguards are, therefore, necessary to protect the persons handling the insecticides, the crops to which they are applied, the consumers who have to use the produce of the treated crops, and the domesticated animals which feed on such produce, the pollinating insects, the parasites and predators of insects, the fish, birds and other wildlife preying upon insects, likely to be affected by coming into direct or indirect contact with the insecticides applied. Also, the pollution of air, soil and water is to be avoided as far as possible.

Some of the basic precautions to be taken while handling, using and storing insecticides are as follows:

1. Keep the insecticides in closed, properly labelled containers in a dry and cool place, away from food and fodder, and in places where children and animals cannot reach them.
2. Use the insecticides according to the instructions given on the containers, and adhere to the dosages recommended.
3. Persons handling insecticides should avoid the contact of the insecticides with their skin (specially that of the insecticides listed as toxic and highly toxic) and the inhalation of dusts, vapours or mists. The minimum precautions of wearing rubber gloves and covering the eyes and nose should be taken. The operators must not smoke, eat or drink anything while applying insecticides. After finishing their work, they should take bath, wash their hands and face with soap thoroughly and change their clothes. The clothes worn by them during the operations should be washed properly.

Application Equipment for Insecticides: The proper choice of the equipment to apply insecticides depends on a number of factors, the first of which is the formulation to be used. Insecticides can be applied as dusts, sprays (as high-volume and low-volume sprays, aerosols, mists or fog), or as granules, and the application machinery has to be chosen accordingly. The next important factor to be considered is the size of the area to be treated. For small holdings, hand-operated applicators may be suitable, whereas for large holdings power-operated machines will be needed. Similarly, due consideration is to be given to the crop to be treated and the pest to be controlled.

Different kinds of dusters and sprayers available in the market are listed below.

Dust Particles: Manually-operated dusters for small holdings. These are plunger-type bellows duster, knapsack duster, and crank or rotary duster (very commonly used in India).

Dust Power: These dusters are used to cover large areas and for quick coverage. These are engine and traction-operated.

To obtain good results and for proper coverage, a duster should produce a continuous and uniform cloud of the insecticide. The dust should be dry and fine enough not to clog the hopper or lump inside it. The duster should be light and portable and should be easily repaired even by the village blacksmith.

Different Sprayers

Sprayers used by Man: These are hand-sprayers or atomiser (for treating a few plants only), stirrup pump or sprayers, slide-pump sprayer, knapsack sprayer (pneumatic or battery-operated), compressed-air sprayer, rocker sprayer, foot or pedal sprayer and wheelbarrow sprayer. Pneumatic knapsack and foot or pedal-pump sprayers are commonly used by the Indian farmers with small holdings.

The power-operated sprayers include the equipment operated with gasoline engines, such as stretcher sprayers, wheelbarrow sprayers, boom sprayers and self-propelled high-clearance sprayers. Power-sprayers are generally available with village panchayats or zila parishads, from which the farmers can obtain them on hire, whenever necessary.

In addition, there are special types of machines for producing mists, fogs and aerosols which have limited use. Recently, granule applicators have been marketed for applying granular insecticides.

The proper care of application equipment is of utmost importance. The machines should be kept in good working order even when not required immediately. They should be cleaned thoroughly, oiled and greased before storing after use.

In the case of sprayers, insecticides should be completely removed from the tank and the delivery tube and properly dried; their nozzles should be cleaned and any leaks should be immediately repaired.

Use, How Much Insecticides: For controlling the insect pests of crops, the usual practice is to indicate the dosage, i.e. the quantity of the active ingredient to be applied per unit area (acre or hectare). However, the dosage may have to be varied, depending on the type and the stage of the crop. Another method is to indicate the concentration and the quantity of an insecticide to be dustered or sprayed. Here also, the quantity of the dust or of the spray liquid has to be adjusted according to the type and the growth stage of the crop and according to the application equipment available. In the following pages, concentrations of insecticides have been recommended, keeping in view the general use of high-volumes prayers, by which the plants may be thoroughly covered up to the run-off stage. The quantity of the spray may, therefore, have to be adjusted accordingly.

Effect on Crops

The crops grown in India are attacked by a large number of different pests, some of which are specific to a particular crop or feed on related species of a particular family, or to crops grown in a particular season. However, there are other insects, such as termites, army worms, cutworms, grasshoppers, and locusts which attack a large variety of crops in different seasons. Out of them, locusts occupy a unique position because of their swarming habit and migrations to distant areas irrespective of international boundaries. An account of this group of insects, specially the desert locust, which is regarded as 'Farmers' Enemy No. 1', is given separately.

Important Locusts: India has been subjected, from time immemorial, to locust invasions which cause immense damage to crops and trees, sometimes resulting in severe fodder famines and considerable loss of livestock.

General Features: Locusts are closely allied to short-horned grasshoppers from which they differ markedly in certain respects. Both are anatomically very similar, feed on vegetations, lay eggs in the soil and pass through several instars before they become adults. Whereas grasshoppers feed on specific crops only, locusts devour almost all plants and vegetation, and even feed upon dead and dying locusts, i.e. they are omnivorous. Unlike grasshoppers, locusts exist in two phases: (*i*) the solitary phase, and (*ii*) the gregarious phase. In the solitary phase, they are inactive, live and breed as scattered individuals, and are relatively harmless. In the gregarious phase, they are very active, acquire the habit of congregating and forming bands, breed rapidly, and in the adult stage fly long distance during migration even overseas. Climate, weather, soil, vegetation and their mass destruction organized by man cause the transformation of the gregarious phase into the solitary phase.

Types of Locust: There are eleven main species of locust, of which three, viz. the Bombay locust, the migratory locust and the desert locust, are found in India. The first two species live and breed in the country and are found scattered all over western India from Rajasthan in the north to Karnataka and Tamil Nadu in the south. They rarely form swarms. They mainly feed on plants of the grass family and occasionally when they become gregarious, do considerable damage to *jowar, bajra* and other millets.

Desert Locust Distribution: The desert locust is the most destructive international pest, its distribution extending from India in the east to Africa in the west, passing through Pakistan, Afghanistan, Iran, Iraq, southern USSR, Arabia and in the north, east and west of Africa. In this vast region, the locust lives and breeds endemically, forming swarms and undertaking long flights to distant places. The locust invasions occur in cycles, the duration of a cycle as well as of the period between two cycles being variable.

Different Phases: In the solitary phase, the locust is more or less harmless, whereas in the gregarious phase, when its hoppers congregate and march in bands, destroying all vegetation in their path and give rise to swarms, it is very destructive. The hoppers of the solitary phase are green throughout their life, but the hoppers of the gregaric is phase are black in the first two stages of development and their dark colour turns greenish yellow and orange in the later stages. The adult locust of the solitary phase remains grey, but the adult is pink at first, but turns grey and yellow as it matures.

Breeding Time: There are generally two breeding seasons in a year: (i) winter-spring, and (ii) summer-monsoon. In regions where the rainfall is mainly received in winter and early spring, e.g. in south-east Arabia, southern Iran and Baluchistan, the locust breeds in spring. In India and other countries where rainfall is received during the summer months, breeding takes place at that time of the year. The swarms produced during winter and spring in the Middle East and Baluchistan usually fly eastwards into Pakistan and India, in summer. With the advent of the monsoon in this area, they lay eggs from June to September. The swarms resulting from this breeding return during autumn to the areas of winter rainfall when conditions in the summer-rainfall area become dry and unsuitable for breeding. A number of swarms developed in the country fly also to the north, south and east, and thus invade all parts of India and damage the *kharif* crops.

Sometimes, breeding takes place in winter, specially in Punjab and Haryana when they damage the *rabi* crops as well.

History of Growth: The female of the gregarious phase lays eggs in clusters of 40-120 in a hole in moist soil to a depth of 75 to 150 mm and covers them with a liquid material that soon hardens to form a protective covering. Young hoppers hatch from the eggs in two to three weeks in summer, and in three to four weeks in autumn and spring. There are five stages in the development of hoppers. At each stage they cast off the skin every sixth or seventh day as they progressively become bigger. As long as they are hoppers, they have only rudimentary wings. After the fifth moulting, they develop wings from the wing-pads present in the late hopper stages. During summer and monsoon, the hoppers complete their development in four to five weeks; and during autumn and spring in six to eight weeks. As the hoppers moult, they undergo a change in colour and size. In early stages, they are almost wholly black, and in the later stages, they are black, with orange and yellow patterns. When they acquire wings, they are pink and ultimately turn yellow, when fully mature.

Both hoppers and adults feed on all types of vegetation, though there are some plants which they would not touch, or eat only with reluctance.

Locust Cycles and Biotic Theory of Periodicity: The theory has been recently propounded to explain the causes of periodic locust outbreaks. According to this theory, the actually adverse climatic conditions of the desert are, in fact, favourable to the locust, as they adversely affect its enemies (specially cold-blooded reptiles), which are responsible for not allowing the locust to breed outside the desert, where the conditions are otherwise very favourable. The desert environment becomes more inhospitable to the enemies of the locust during the minima of the sun-spot cycle (periods of drier and more intense heat), leading to a reduction in the population of the locust enemies which consequently leads to an increase in locust population and swarm formation. During the maxima of the sun-spot cycle (period of less dry and less intense heat) on the other hand, the desert environment becomes favourable to the enemies of locusts which migrate to the interior of the desert from the periphery and bring down the locust population, thus breaking the locust cycle.

This biotic theory has made it possible to visualize the possibilities of checking the start of the locust cycle in the early stages by creating suitable ecological niches in the locust-breeding areas, so that their enemies can migrate and survive the adverse conditions in such areas and continue to keep down the locust population.

Measures of Control: There are two main methods of locust control, viz. *(i)* the use of poison baits, and *(ii)* dusting and spraying with insecticides. Poison-baiting, though cheap, is not very practicable, dusting and spraying have been found to be very effective. In the egg stage ploughing or flooding the land has been practised in the past. This method is not very effective inasmuch as it destroys only about 30 per cent of the eggs and renders subsequent control measures with insecticides difficult. Spraying the egg-infested, uncultivated land with Aldrin is very effective, as it leaves a poisonous film on the surface with which the emerging hoppers come into contact and die. Once the hoppers have emerged, they can be effectively controlled by dusting or spraying or both, as the hopper stage is most vulnerable. For dusting, BHC is the most effective insecticide. For young hoppers, 3 to 5% dust is sufficient to kill them; for advanced stage, 5 to 7% dust is necessary; and adults need 10% BHC. Dusting or spraying may be done on the hoppers directly or it may be done on strips of ground, two to three metres wide, in the path of the advancing hoppers. Another method is to dig trenches 45 to 75 cm deep and 30 to 60 cm wide, depending on the size of the hoppers, in front of the moving hoppers to which they may be driven and buried alive. The dusting or burning of bushes in which hoppers rest can also be resorted to.

Adult locusts, flying in swarms, are most difficult to control. If they alight in any field or in a piece of ground try to kill as many of them as possible, by beating or crushing, specially in the morning when they are lazy. Burning them by setting fire to dry bushes on wastelands, or by flame throwers, if they have settled on green bushes, hedges or trees if useful. This method works well if the locusts have settled for resting, mating or egg-laying. Poison baits-spread at the rate of 20 to 50 kg per hectare are useful if locusts have settled to feed in cropped feeds. If the swarm consists of pink locusts, dust them with 10% BHC. Aerial spraying of the flying swarms and ground hoppers has to be resorted to with Aldrin solution in oil in the case of very serious infestation. However, this cannot be done when the swarms and hoppers are in the populated areas.

Despite the discovery of a large number of potent insecticides, which can be used against locust, both from ground and air, the farmer has been finding himself to be only a helpless onlooker once the locust swarm settles in his field. Though part of the swarm may be killed, he invariably loses his crop.

Situation has materially changed recently after the discovery of the anti-feeding property of the *neem* kernels. It has been found that 0.1 per cent suspension *of neem* kernels, sprayed on the crop affords protection against locust feeding for about 14 days. The cultivator can locally collect and preserve the *neem* seeds till required. By spraying his crops with a *neem*-kernel suspension, when the invasion of his fields, by hopper bands or swarms is imminent a cultivator can now ensure the safety of his crops. Simultaneously, the control operations can be carried out by adopting suitable methods.

Organizational Structure: As the locust is an international pest, control measures are effective only when they are properly organized at the national and international levels. Since 1951, the FAO has taken good interest in the desert-locust control by periodically appraising the overall situation, recommending suitable control techniques and organizational patterns, and by coordinating and contributing to the carrying out of control campaigns in certain key areas. More recently, it has been organizing international campaigns annually in the Arabian Peninsula with the assistance of a number of countries.

In India, the Central Government has assumed direct responsibility for controlling locusts in the desert areas of Rajasthan, Punjab, Haryana and Gujarat. This area is divided into four circles with headquarters at Jodhpur, Barmer, Bikaner and Palampur, each of which is subdivided into zones and the zones are further divided into outposts. In each zone and at each outpost, technical staff, control equipment, vehicles, etc. are maintained. Wireless-sets are used for transmitting messages and aeroplanes are available for aerial spraying and dusting.

In the cultivated or non-desert areas, the responsibility for controlling locust rests with the state governments. The State Plant Protection Organization provides the necessary technical staff, equipment, vehicles, etc. The District Magistrate is in charge of locust-control operations, ill carrying out which, he is helped by the State Plant. Protection Officer and the district staff and extension workers of the Revenue and Agricultural Departments.

In locust control, the exact information on mating, egg-laying, hopper concentration, swarm movements, direction of their flights, the size of the locusts, their colour, etc. is of utmost importance for the rapid transport of the necessary equipment and insecticides, the posting of the technical staff, the supply of labour, etc. All people, both officials and non-officials, can therefore, render help by the prompt transmission of the information of the kind mentioned above; they may be having or may have heard to the nearest authorities concerned, e.g. any officer of the Revenue or Agricultural Department or of Community Projects or National Extension Service.

About other major pests of different crops and their control, information in a condensed form is given in the following pages.

Various Diseases

How to Define: A plant disease is a physiological or structural abnormality that is deleterious to the plant or to any of its parts or products, or reduces their economic value. It expresses itself in a definite series of symptoms.

Nature of Disease: Plant diseases may be temporary or permanent, depending on the speed of recovery. Further, a disease may be localized, involving only a limited part of a plant, or it may be systemic when its infection is generalized and spreads throughout the plant body.

Causes of Disease: Plant diseases may be caused by micro-organisms, including fungi, bacteria, viruses, mycoplasmas, etc. or may be incited by physiological causes including high or low temperatures, lack or excess of soil moisture and aeration, deficiency or excess of plant nutrients, soil acidity or alkalinity, etc. The micro-organisms obtain their food either by breaking down dead plant and animal remains (saprophytes) or by attacking living plants and animals (parasites). In order to obtain nutrients, the parasitic organisms excrete enzymes or toxins and kill the cells of the tissues of the host plant, as a result of which either the whole plant or a part of it is damaged or killed, or considerable disturbance takes place in its normal metabolic processes.

How to Reproduce: Micro-organisms are reproduced and disseminated mainly by spores. Wind, water, seed, diseased plants, cuttings and tubers, animals, men and insects, and soil help to disseminate diseases. Virus diseases are spread as the result of direct contact, through seeds, diseased cuttings and grafts, and more commonly by insect and other vectors. Most of the fungi and bacteria errupt or ooze from the surface after the host tissues have been damaged, whereas the viruses work altogether internally.

Therefore, the control of plant diseases usually lies in preventing infection, as it is very difficult to kill the fungi or bacteria or to inactivate the viruses inside the host plants. However, as the result of recent studies made in the development of systemic fungicides, it appears that in the near future the diseased plants may even be cured. The revolvement of disease-resistant varieties is another important and cheap method of controlling diseases. However, this method of control has its limitations because of the appearance of new races of the pathogens from time to time.

Disease Borne by Seed Itself: In the case of seed-borne diseases, the pathogens are carried either on the surface of the seed or within it. When a pathogen is external, it may be destroyed by treating the seed with a chemical, e.g. formalin, TMTD, copper carbonate, Captan. organo-mercurials, e.g. Agrosan and Ceresan. Except Formalin, all chemicals are available in the dust form. The recommended dust is thoroughly mixed with the seed, preferably in a rotating-drum, so as to uniformly cover the seed surface with a layer of it. While using formalin, dilute it with an equal quantity of water and spray it over a thinly spread layer of grains. The grains are then heaped and kept covered with a tarpaulin overnight. For internally seed-borne diseases, hot-water treatment and solar treatment are very efficacious.

For the treatment of wheat seed with hot water to control loose smut, the seed is soaked in ordinary water for four to five hours and then dipped into hot water at 49° C for one minute and then immersed again into hot water at 54° C for ten minutes. In the case of solar treatment, which is carried out on a bright sunny day during the summer months, the seed is soaked in water for 4 hours from 8 a.m. to 12 noon. It is then taken out and immediately spread in a thin layer on sheets of cloth to dry in the sun. Recently, systemic organic compounds have also been shown to be effective in eradicating both externally and internally seed-borne infections.

Disease Borne by Soil: In the case of soil-borne diseases which attack the seeds or seedlings, the use of soil disinfectants or sterilizers is helpful in destroying the pathogens. The soil treatment involves the use of chemicals formaldehyde,

Captan, Thiram, Zineb, organo-mercurials, PCNB, ethylene dibromide, Vapam, etc. They are used to drench the soil, or are broadcast on it, or are applied to it in furrows, or are used to fumigate it.

Disease Born by Air: In the case of diseases caused by air-borne pathogens, the application of chemicals to the foliage is the most effective way of controlling them. Though many fungal and bacterial diseases can be controlled in this way, chemicals have not yet proved very successful in controlling virus diseases. In their case, protective measures have to be aimed at controlling the insect vectors by spraying insecticides on the host plants or by using disease-free material for propagation.

Fungicides: Copper, sulphur, organo-mercury and organo-sulphur compounds and antibiotics are some of the important fungicides. Recently, some systemic organic compounds have also been developed. The choice of fungicidal compound depends on the nature of the disease and on the susceptibility of the crop. They are available as dusts or wettable powders and sometimes in liquid form. The method of applying them depends on the availability of water, suitable machinery, the nature of the crop to be treated, the area to be covered and the cost of labour.

Fungicides with Copper: Bordeaux and Burgundy mixtures, and the dispersible forms of cuprous oxide, copper oxychloride, and basic carbonate are among the important copper fungicides. Bordeaux mixture is prepared by mixing solutions of copper sulphate and quicklime in certain proportions. For preparing 5:5:50 Bordeaux mixture, 2 kg of copper sulphate and 2 kg of quicklime are dissolved in 225 litres of water. At first, the stock solutions of the two constituents (1/2 kg in 4 litres of water) are prepared. Copper sulphate is dissolved by placing it in a piece of gunny sacking and then suspending it in water in an earthen or wooden vessel.

Lime is first slaked with a small quantity of water and is made into a thin paste before the remaining water is added to it. The two stock solutions are mixed together just before use by pouring the copper sulphate solution into the lime solution. The final reaction of the mixture should be either neutral or slightly alkaline. The reaction can be tested with litmus paper, or by dipping a bright iron nail into the mixture. The litmus paper must not turn red and the nail must remain bright, i.e. without any copper deposit on it. Bordeaux mixture deteriorates on standing; hence only fresh preparations should be used. However, this deterioration can be prevented appreciably by adding cane-sugar to the mixture at the rate of 60 grammes per 225 litres.

Burgundy mixture is a variation of Bordeaux mixture, in which lime is replaced by sodium carbonate (washing-soda). The method of preparation is just the same as for Bordeaux mixture. The usual strength of the solution is 4—6¼—50. There are a number of proprietary copper fungicides available in the market, e.g. Perenox, Persian, Blitox, Cuprokyt, Cuprosan, Fytolan, which are as efficacious as, and in some cases better than, Bordeaux mixture. The proprietary products also contain dispersing and wetting agents which make them more effective.

Fungicides with Sulphur: The important sulphur fungicides in use are sulphur powder, wettable sulphur, lime-sulphur and organo-sulphur. Sulphur-dust is a very cheap fungicide, but its efficacy depends upon the size of the particles which should pass through a sieve of 250 to 300 meshes per inch. Wettable sulphur is used on crops which are liable to suffer from lime-sulphur scorching. The lime-sulphur solution is prepared as under:

Water (225 litres) is first boiled in a vessel of iron, tin or zinc (but not of copper) and then quicklime (20 kg) and sulphur (55 kg) are added to the boiling water. The mixture is stirred and boiled for one hour. The resultant liquid is strained and its density is determined with a hydrometer ; enough water is added to bring the specific gravity to 1.005 or 1.010 (see table, p. 300) before use. Proprietary products, e.g. Sulfinette and Limsul, are available in powder form. On dilution, they are ready for use as the lime-sulphur spray. The lime-sulphur spray should not be used on tender foliage, as it is likely to scorch it. A number of organo-sulphur compounds, e.g. thiram and dithiocarbamates, are extremely safe to use on most of the plants with tender and sensitive foliage. Tulisan, Zineb and Ziram are some of the proprietary products, based on organo-sulphur compounds.

Dilution Table for Lime-sulphur

Density (Degree Baume)	*Gallons of water to be added to one gallon of concentrated solution to give:*	
	Sp. gr. 1.005	*Sp. gr. 1.010*
28	22	46
29	23	48
30	24	50
31	26	52
32	27	54
33	28	57
34	29	59
35	30	61
36	31	63

(a) *The Organo-mercurials:* These have proved to be effective in controlling seed-borne and some soil-borne diseases. They are applied as dusts in extremely small doses to the seed before sowing. They have no harmful effect on germination. Rather, they prevent emergence losses caused by soil-borne pathogens and result in a better stand of the crop. There are a number of organo-mercury preparations, e.g. Agrosan GN, Ceresan, Semesan and Parrugen. They are, however, poisonous to human beings and animals. Hence great care is necessary in handling these fungicides and the seeds treated with them. Some vegetable seeds and other seeds, e.g. those of linseed, are non-tolerant of mercury. In the recent years certain chemicals, such as Captan, Thiram, Busan-72 and Dithane M-45, and some systemic fungicides have been developed. They may be used for seed treatment instead of the mercurials.

(b) *The Systemic Fungicides:* Systemic fungicides can move mainly upwards only within the plant and kill the pathogen away from the point of application. They can be applied as seed or soil treatment, and as sprays. Some of the important systemic fungicides are: oxathin derivatives (Plantvax and Vitavax), Benlate, Demosan and Bavistin.

(c) *The Antibiotics:* Antibiotics are mainly used against bacterial plant pathogens. Some of the promising antibiotics are: streptomycin sulphate, streptocycline, terramycin, agrimycin-100 (a mixture of streptomycin and tetracycline), agrimycin-500 (containing copper), etc.

Cultural Control: Cultural practices, such as crop rotation, mixed crop ping, adjustment of the date of sowing, depth of sowing and field sanitation through the removal of diseased debris, weeds, alternate and collateral hosts are very effective in controlling plant diseases. A strict enforcement of quarantine, especially in the case of imported plants, helps to prevent the introduction of new diseases into a country.

As virus diseases are difficult to control or cure with chemicals, the use of disease-free propagating material is very important. For this purpose, it is necessary to establish disease-free nurseries at isolated places.

Control measures against some of the important diseases, together with their casual organisms and symptoms, are described in the following pages.

Plant Ailments

Though plant parasitic nematodes often cause severe damage to cultivated crops, their role as a limiting factor in agricultural production has not been generally appreciated. These microscopic organisms, sometimes also called thread worms, eelworms, round worms, etc. mostly inhabit the soil in association with the roots of plants. A few species feed 0 : 1 the aerial parts of plants too, such as stems, leaves, ear heads, grains, etc. Unlike insects and other disease-causing agents like bacteria, fungi and viruses, nematode populations increase rather slowly, and the diseases caused by them are, generally, not of sudden epidemic type resulting in obvious destruction of the crop over vast areas in a matter of days, but cause slow decline in yields spreading over years. Symptoms of nematode diseases may not often be clear-cut and easy to pin-point, and the disease may itself be the result of interaction between nematodes and other pathogens like bacteria, fungi and viruses.

Many estimates have been made of crop losses caused by pests and diseases. However, many of these are often questioned as being unrealistic. Assessment of crop losses due to nematodes is even more difficult as compared to insects for want of precise techniques and adequate survey data. Some workers have attempted to express these losses in terms of money. In the USA alone, the estimated annual loss in cultivated crops due to plant nematodes was considered to range between 250 and 500 m dollars till a few years ago. A Committee on Crop Losses recently appointed by the Society of Nematologists, USA, has undertaken a detailed study on the subject. On the basis of an average 6 per cent loss in field crops, 12 per cent loss in fruit and nut crops, 11 per cent in vegetable crops and

10 per cent in ornamental crops, they estimated the annual losses in farm value suffered by that country to a staggering total of 1,590.7 m dollars. The factors considered in making these estimates included reduced yield, reduced market quality owing to nematode damage, increased yield or quality or both after chemical treatment and cost of chemical control measures. Data from experiments involving extreme damage levels by high nematode populations and greatly accelerated growth rates after successful control were discounted. An assessment in England and Wales, some time ago, indicated that the annual loss of potatoes caused by the golden nematode of potatoes alone was of the order of 200,000 tonnes representing £2,000,000.

In India, some attempts have been made recently to estimate losses due to nematodes in a few crops. Annual losses as high as Rs 40,000,000 in wheat and Rs 30,000,000 in barley due to the *molya* disease caused by *Heterodera avenue* in Rajasthan alone are considered common and reasonably correct. The ear-cockle nematode of wheat has been estimated to cause an annual loss of over Rs 75,000,000 even at a low infestation level of less than 1 per cent. Another disease, a decline of coffee in South India, caused by a root lesion nematode *Pratylenchus Coffee* has been estimated to cost that industry rupees 20,000,000 per year. It is rather premature to attempt estimation of losses in other crops for want of sufficient information.

Nematode Pathogens: Nematode parasites of higher plants are by no means the only nematode fauna of the earth. There are numerous free living forms in the soil, besides parasites inside the bodies of man and animals. Countless numbers inhabit the seas also. Even so, plant parasitic nematodes are so wide-spread and universal in distribution that no living plant is likely to be free from their attack. All cultivated crops are usually attacked by one or more species of these nematodes. The actual damage done to the host plant, however, would depend on the population, the age of the plant at the time of infestation and several other ecological factors.

Plant nematodes are mostly microscopic, generally ranging from 0.5 to 1.0 mm in length. Like insects, they vary in size, appearance and habits. Typically they are cylindrical, the body tapering towards the head and tail-ends. A departure from this typical slender shape is observed in mature females of many species which assume swollen pouch-like, lemon, pear, kidney or other irregular shapes while the male remains slender. Perhaps the most outstanding diagnostic feature of plant parasitic nematodes is that they all have at the mouth end a stylet or spear, which is a sharp pointed tube like a doctor's hypodermic needle. It is worked back and forth into plant tissue with the help of three sets of muscles attached to its base. The plant sap is sucked into the alimentary canal through the stylet tube by the vacuum created by the alternate contraction and relaxation of a median valve situated posterior to the stylet. At the same time, the nematode also injects a small quantity of its oesophageal gland secretion containing digestive enzymes into the plant tissue. The reaction of the plant and the expression of various disease symptoms are attributable to some toxic substances thus injected and to depletion of plant nutrients while feeding. Nematodes have six stages in the life cycle, the egg, four larval stages and the adult. One cycle may be completed in a few days or may require several years depending upon a great variety of conditions and circumstances.

The nematode parasites of plants can be classified in a number of ways on the basis of feeding habits, parts of plant attacked, the type of reproduction, etc. A simple classification based on their parasitic habit is adopted below:

I. *The ectoparasites:* These nematodes are external feeders and normally do not enter the plant tissue. They feed on the cells of small feeder roots near the root cap and may cause death of cells near the feeding site resulting in collapse of tissues. Spiral nema todes *(Helicotylenchus, Rotylenchus)*, lance nematodes *(Hoplolaimus)*, ring nematodes (*Criconemoides, Hemicriconemoides*), stunt nematodes (*Tylenchorhynchus)* etc., are good examples. Some of these ectoparasites may cause small galls on the roots, as for example, sheath nematodes *(Hemicycliophora)* and dagger nematodes *(Xiphinema).* Besides the root feeding forms mentioned above, certain species of foliar nematodes *(Aphelenchoides)* also feed ectoparasitically on the leaves and flower buds. Deformed leaves and fruits or a reduction of these parts may result from heavy infestation.

II. *Semi-endoparasites or exposed endoparasites:* The citrus nematode *(Tylenchulus semipenetrans)* and the reni-form nematode *(Rotylenchulus reniformis)* are two good examples of this category, both widely distributed in India. The adult females are saccate or swollen with the head and neck region alone embedded in the roots and the hind part of the body protruding outside the root tissues. Both species cause appreciable damage to their hosts and are of great economic importance to India. While the citrus nematode is mostly specific to citrus, the reniform nematode infests a variety of crops like millets, pulses, cotton, castor, vegetables, chillies, potato etc., and is one of the most harmful species in the country.

III. *The endoparasites:* These nematodes are completely internal parasites and include some of the most destructive crop pests known to man.

(i) *Nematodes of root-knot (Meloidogyne spp.):* These are the most commoh plant parasitic nematodes in India and their infestations are easily recognised by the characteristic root-galls they cause. Vegetable crops like tomato, brinjal, bhindi, cucurbits, crucifers, pulses, potato, sugarcane, banana, citrus etc., are all highly susceptible hosts. Even cereal and millet crops, which are generally not favoured, are attacked under certain conditions and suffer considerable damage. Within the root-galls are harboured the highly swollen or flask-shaped adult female nematodes which are sedentary in habit, incapable of movement within the roots. More than half a dozen species of root-knot nematodes are known to occur in India infesting various crops. *M. incognita* and *M. javanica* are the most common species.

(ii) *Nemotodes regarding cryst-forming (Heterodera spp., Globodera spp.):* These are also sedentary endoparasites and probably the most highly specialised among the plant nematodes, because of the special adaptation for survival which they possess, namely, the cyst stage in the life-cycle. Eggs are generally not laid outside, but retained within the female body wall itself which eventually gets hardened into the brown, leathery cyst. The cysts enclosing the eggs get dislodged into the soil and are very resistant to desiccation and other adverse conditions. It is mainly this adaptation which makes the cyst-forming nematode the most difficult to control. *G. rostochiensis* and *G. pallida* on potatoes in the Nilgiris, *H. avende* on wheat and barley in north-western India, *H. cajani* on pulses and *H. zeae* on maize are good examples.

(iii) *Nemotodes regarding root-lesion (Pratylenchus, Hirschmanniella, Radopholus):* These are internal root parasites whose adults are migratory in habit and move about through the cortical tissues of the root. Extensive injuries in the form of root lesions are generally caused, often with severe symptoms. Host plants include many important cultivated crops such as rice, vegetables, millets, pulses, cotton, sugarcane, bananas, citrus, coffee and many others. The root-lesion nematodes *Pratylenchus thornei, P. zeae, P. delattrei, P. coffeae* etc., on a variety of crops; the burrowing nematode *Radopholus similis* on banana, peppervine and coconut in South India ;

and the rice-root nematode *Hirschmanniella oryzae* are more important species.

(iv) *Parasites in flower and seed:* Under this category are included the white tip nematode of rice, *Aphelenchoides besseyi,* which is a seed-transmitted nematode causing distortion and chaffiness in rice grains, and the wheat gall nematode, *Anguina tritici* which produces galls in the flowering parts of the host plant. These two species are known to cause wide-spread damage to rice and wheat respectively in India.

(v) *Nematodes regarding foliar (Aphelenchoides spp.):* These live as internal parasites of buds and foliage of plants like chrysanthemum, strawberry, etc., and may cause distortion and crinkling of foliage, and leaf spots. At least one species *A. ritzemabosi* is known to infest ornamental crops in Simla and other hill areas.

(vi) *Nemotodes regarding stem and bulb (Ditylenchus spp.):* These occur more commonly in temperate regions of North America and Europe. *D. dipsaci* infests onion, garlic, alfalfa, clover, bulbs, ornamentals etc., in these countries. *D. destructor* causes dry rot of potatoes. There is no published record to indicate the presence of either species in India.

Plant Injury: Plant parasitic nematodes are obligate parasites. During feeding in or on the roots, they penetrate cells and inject their salivary fluid into the host tissues. It is apparent that a number of digestive enzymes like carbohydrates and proteases are released during the process of feeding resulting in hydrolysis of host components and leading to altered host metabolism. Feeding on plant cells by nematodes may deprive the host of nutrients and water necessary for its growth. The conductive tissues responsible for the translocation of nutrients and water to the tops are often blocked.

Metabolites vital for the growth and maintenance of host tissues are often blocked at the site of infestation. Feeding injury will thus impair the efficient functioning of the organs concerned. The impact of mechanical injury to the host tissue brought about during the process of feeding or penetration by enormous number of nematodes has also significant effect on plant growth.

Many plant parasitic nematodes, e.g. *Meloidogyne* spp., *Rotylenchulus reniformis* etc., may be directly responsible for poor germination and thin seedling stand due to pre and post-emergence seedling damage under field conditions. Root-knot and reniform nematodes are also known to interfere with rhizobial nodulation in leguminous crops leading to reduction in the number and size of nodules and consequently reduced nitrogen fixation. Often nematode injury leaves avenues for the entry of other secondary organisms. The association of plant parasitic nematodes with other microorganisms, like fungi and bacteria in disease complexes, is now a well recognised phenomenon and very often such association increases the severity of the disease. Further, the altered metabolism induced by nematode parasitism makes the plant vulnerable to the attack of otherwise weak pathogens. The discovery of some of the plant parasitic nematodes to be the vectors of soil-borne virus diseases indicates further potentiality of these small animals to cause injury to crops.

Damage and Symptoms: The symptoms of nematode injury to host plants will vary with the species of nematodes involved, initial population density, age of the plants and many other ecological factors. Loss of vigour of individual plants, patches of stunted plants, general decline followed by reduction in crop yield etc., are some of the common symptoms. Field symptoms are often of a generalised type, unspecific, giving an impression that the crop is suffering from shortage of water and fertilizers even though both these factors are available in plenty in the soil. General chlorosis, die-back, leaf drop, small-sized foliage, dead buds, leaf distortions, under-sized and premature ripening of fruits, root galls, root rot, root lesions, excessive proliferation of roots or devitalized root tips are some of the symptoms of nematode injury to the hosts. Since symptoms on host due to nematode infestation are similar to those caused by other pathogens like bacteria, fungi, etc., laboratory examination of soil and root is necessary to make sure of the role of the nematode in the manifestation of the disease.

Host and Parasite: Among all the known pathogenic entities, the kind of relationships that exist between plant parasitic nematodes and their hosts can be considered to be rather unique, perhaps in the larger sense only just a degree below symbiosis. Nematode infestation seldom results in outright killing of the host with populations as observed under field conditions. A very low initial inoculum level is known to have even stimulatory effects on the host plants.

The degree of parasitism can be considered to be most advanced when the nematodes stimulate changes in the host cells vital to their own growth and development, without the destruction of the infested cells. The citrus nematode, *Tylenchulus semipenetrans* establishes a group of six to ten 'nurse cells' in the cortex rich in metabolites. Similarly the cyst-forming (*Heterodera* spp., *Globodera* spp.) and the root-knot nematodes (*Meloidogyne* spp.) produce a group of 'giant cells' around their head end in the stelar region of the hosts. In addition, the root-knot nematode induces extensive pericycle hyperplasia and cortical hypertrophy which result in the formation of galls. In contrast to these physiologically specialised nematodes, the activity of less specialised nematodes may affect the host cells drastically. Removal of cell contents following stylet penetration usually results in collapse of tissues and death of cells.

The deleterious effect on hosts due to nematode infestation is, by and large, physiological. Such physiological changes in hosts are reflected in increased respiration, accelerated enzymic activity of oxido-reductase and hydrolases, increased protein,

DNA and RNA synthesis and changed rate of photosynthesis. The exact mechanism how plant parasitic nematodes induce physiological imbalance in the host is rather intriguing and remains unanswered. However, evidence is fairly strong to suggest involvement of plant growth regulatory substances as in the case of the root-knot disease. A number of auxins like indole acetic acid, indole butyric acid, indole acetonitrile and indole acetic acid ethylester are reported from different species of *Meloidogyne.* It is possibleth at auxins of nematode origin are directly injected into the host tissue to bring about disturbed metabolism and cellular abnormality. Alternatively, it has been suggested that proteolytic activity from root-knot nematode larvae may break down plant proteins, subsequently leading to the accumulation of IAA, and this may result in the formation of galls. On the other hand, resistance of host plants to nematode invasion is characterised by hypersensitive reactions, failure of the parasite to establish permanent feeding sites (giant cells in *Heterodera, Globodera, Meloidogyne,* and nurse cells in *T. semipenetrans)* and thickening of cell wall around nematodes head. Differences in phenolic substrates in the hosts or in B-glucosidases in the parasites have been accounted for resistance. Fate of nematodes in a resistant plant is reflected in retarded rate of reproduction, delayed maturity of the parasite, production of more males and failure of nematode to complete its development.

Control Method: All control methods are directed to bring down the nematode population low enough to raise the crop profitably. The choice of a particular method will primarily depend upon the nematode species and the crops involved. The following is a brief outline of different control methods against plant parasitic nematodes.

Method Belonging to Culture: The objective of these measures is to keep a field free of susceptible plants or adopt suitable land management practices so that nematodes will be reduced to a very small number. This includes crop rotation, fallowing and ploughing during the summer months, mulching and use of resistant or tolerant varieties.

(a) *Rotation of Crop:* This is one of the recognised methods of keeping nematode populations under check, especially where they have distinct host preferences, e.g. in the case of cyst-nematodes of potatoes, wheat, barley, etc. Susceptible crops should be grown only once in a few years, rotating them with highly resistant varieties or non- host crops. Vegetable crops should not be grown repeatedly in the same field, but should be rotated with cereal crops in order to reduce infestation by root-knot nematodes.

(b) *Summer Fallow and Ploughing:* Hot dry summer conditions prevalent in India could be made use of for nematode control with very satisfactory results. Nematodes are very sensitive to heat and the drying action of the sun and the wind. Keeping the land fallow during summer months and ploughing the soil deep 2-3 times at intervals of 10-15 days is an excellent method to reduce root-knot nematodes and others to the minimum.

(c) *Use of Organic Amendments:* Soil amendment with green manure, compost, oil-cakes (of *neem mahua,* mustard, groundnut, cotton, linseed, *karanj),* saw dust, etc. has been found to reduce nematode populations. *Neem, karanj* or groundnut cakes incorporated into soil at the rate of 1,500 kg/ha gives good control of root-knot and other plant nematodes and could be practised wherever possible. Apart from encouraging the multiplication of natural enemies like nematode-trapping fungi, the decomposition products of these organic amendments are toxic to nematodes.

(d) *Resistant use and Tolerant Varieties:* This is one of the most practical methods of controlling nematodes and most economical from the farmers' point of view. However, the rare availability of good sources of resistance, the long period required to evolve a commercially acceptable resistant variety, and the possible evolution of resistance-breaking bio types in some species of nematodes are some of the practical difficulties encountered in adopting this method. Recent experiences emphasise the use of polygenic and combined resistance against nematodes and other pathogenic microorganisms like fungi since both nematodes and fungi are known to be intimately associated in disease complexes. A full and long range collaboration between plant breeders, nematologists and plant pathologists is essential for successful utilization of this method against phytophagous nematodes.

A number of resistant varieties of plants have been evolved by workers in different parts of the world against specific nematodes. These include tobacco varieties 'NC 95', 'Coker 254' and 'Coker 258'; cotton varieties 'Darwinii', 'Auburn 56', 'Dixie King 11'; soyabean varieties 'Habro', 'Jackson', 'Hardee'; alfalfa varieties 'African Moapa', 'India'; cowpea varieties 'Iron', 'Clay', 'Grand'; lima bean variety 'Nemagreen'; capsicum varieties 'Red chilli,' 'Mississippi', 'Nemaheart'; tomato varieties 'Hawaii 5229,' 'Nematex', 'VEN 8' ; sweet potato variety 'Nemagold', etc., against the root-knot nematode *M. incognita*. The release of the resistant tomato variety 'SL 120' by the, Indian Agricultural Research Institute against *M. incognita* deserves special mention here. Likewise, the soyabean variety 'Peking' is reported to be resistant against the cyst nematode. *Heterodera glycines* and the reniform nematode *Rotylenchulus reniformis*, and potato varieties 'Antinema', 'Cobra', 'Ulster Glade' etc., against some biotypes of the golden nematode of potatoes, *Globodera rostochiensis*. Sustained efforts to screen wheat and barley germplasm against the cereal cyst nematode, *Heterodera avenge* is likely to yield fruitful results in the near future. Some of the barley varieties like 'Morocco' and 'Moroccaine' have already shown great promise as sources of resistance against this nematode.

Biological Control Method: Plant parasitic nematodes have their own natural enemies including predatory nematodes and fungi, protozoa, arthropods and viruses, and these offer as possible biological agents to control obnoxious nematodes. To date, such control measures remain of academic interest and very little progress has been made in this direction.

Physical Control Method: The most commonly used physical methods of nematode control are steam sterilization of soil and hot water treatment of planting material. Hot water treatment is widely used for transplanted crops. Effective times and temperatures vary from 240 minutes at 43.3°C to 1 minute at 54.4°C. Usually dormant bulb material like narcissus, iris, chrysanthemum, ester lily, and root stocks like citrus, peach etc., are treated against nematode parasites in some of the western countries. Hot water treatment of rice seed at 52°-55°C for 10-15 minutes is a very effective method to control the white tip nematode, *Aphelenchoides besseyi,* which is transmitted through seed. The control of the ear cockle and 'tundu' diseases of wheat by removal of nematode galls from seed material by dry cleaning or floatation techniques is another example where physical methods have given excellent results all over the world.

Chemical Control Method: This is by far the most effective method to control plant parasitic nematodes. The most successful nematicides are soil fumigants. They may be applied directly as liquids, emulsions or granules, but the active ingredient spreads in the soil in the form of a gas. The halogenated hydrocarbons D-D (a mixture of 1,3-dichloropropene and 1,2-dichloropropane), EDB (1,2-dibromoethane), MBr (methyl bromide) and DBCP (1, 2-dibromo-3 chloropropane) are among the most commonly used soil fumigants, marketed under different trade names by manufacturers. Several other allied formulations are also available. For successful results, it is essential that the soil to be treated is well prepared to a seedbed condition, soil moisture is moderate and uniform, and the soil temperature is between 15°-30°C at the time of application.

Soil fumigants are injected into the soil at a depth of 20-25 cm and at a spacing of about 30 cm. Applicators of various types are in use in many countries, ranging from hand-injectors suitable for home gardens to large tractor-drawn multi-chisel machines capable of covering 2-4 hectares a day. Soil fumigants like D-D, EDB etc., should be applied at least 3-4 weeks prior to planting as they are highly phytotoxic. However, DBCP can be used in irrigation water around the roots of living plants as they are well tolerated by most crop plants at the dosages generally recommended.

A number of organo-phosphorus derivatives and carbamates also possess nematicidal properties to varying degrees and many of them are systemic in action. Some of them are widely used as insecticides also. Phorate, Fensulfothion, Thionazin, Dichlofenthion, Disulfoton, Diazinon, Fenamiphos etc., among the organophosphates, and Al-dicarb, Carbofuran, Methomyl and Vydate among the carbamates have been widely tested against nematodes all over the world. They are generally characterized by being highly active, low in fumigant activity and by their non-persistence. One of the great advantages of these compounds is that they can be applied as granules in the row or in the planting holes at the time of planting. Bare root dips of transplants and seed treatments are other possibilities still in the experimental stage, though there may be phytotoxicity hazards under certain circumstances. Vydate is perhaps the first chemical which has given good nematode control when used as a foliar spray, but has not become popular owing to its high mammalian toxicity and short duration of the protection achieved.

Detection of nematode problems and adoption of appropriate control methods at the proper time is one of the modern methods of increasing the yields of many of our agricultural and horticultural crops. In general it may be stated that chemical control of nematodes has not made much headway in India due to the prohibitive cost of nematicides and the difficulty in application over large areas. In spite of the high cost, a judicious use of nematicides would be economical in commercial crops like cotton, tobacco, sugarcane, grapevine, citrus, vegetables etc., in heavily infested fields where lack of such control would mean very poor or no crops. In many situations an integrated approach involving more than one method like crop rotation, deep summer ploughing, chemical treatment of nursery beds or bare root dips of transplants, and use of resistant or tolerant varieties offer the best promise.

A list of common nematode problems in our important crops and their control is given below.

2

PROTECTING CROPS

DIFFERENT INSECTS

Polyphagous Pests

LOCUST (Literal meaning of the word is 'Plague')

This pest is of International importance and the following three species are commonly found in India ; of which desert locust is predominant all over the country.

(a) Desert locust (*Schistocerca gregaria* Forsk.)

(b) Indian migratory locust (*Locusta migratoria* Linn.)

(c) Bombay locust (*Patanga succincta* Linn.)

Systematic Position

Order : Orthoptera

Family : Acrididae

Host Plants: The locusts are the polyphagous pest and feed on all varieties of agricultural crops, forests, fodders etc. They do not feed on ak, neem, jamun and sesum.

Distribution: This noxious pest is widely distributed throughout the world, affecting various parts of the important continents, like Asia, Africa and Europe. In India the desert locust is found almost around the year in the desert areas of Rajasthan as well as in the sandy parts of Punjab and Gujarat.

Economic Importance and Damage: The locust swarms cause the collosal losses to a variety of vegetation even including the bark of trees. It is well known fact that in the past locust attack resulted in famines in several parts of the world. The outbreak of locust generally depends on the climatic conditions. A perusal of records have shown that locust outbreaks in India occurred in the year 1812, 1821, 1834, 1843, 1869, 1889-97, 1901-3, 1906-7, 1912-15, 1926-31, 1940-46, 1951-55, 1963, and 1993.

Life History: The desert locust passes through three stages in the course of its life cycle, *viz.*, egg, nymph and adult.

Egg: The eggs are laid in batches of 60 to 150 at a depth of 8 to 14 cm in moist sandy or loamy soil. After laying eggs, the female closes the mouth of the hole with a frothy liquid which ultimately hardens into a water proof plug. Generally, breeding starts with the onset of good monsoon showers. A female may lay about 500 eggs in her whole life span. The eggs are yellow in colour and resemble to rice grain. It is about 78 mm long and 1 mm girth. They hatch in 12-14 days in summer and 4 weeks during winter. At this stage a suitable amount of soil moisture is most essential for the proper development of eggs. If this prerequisite is provided by timely rainfall the eggs hatch into a vermiform larvae with membranous covering in which they are enveloped and they enter in the hopper stage of their life.

Nymph (Hopper): The colour, size and habit of hoppers depend upon the phase of locust. Generally, there are two types of phases namely ; solitary and gregarious.

Solitary Phase: In this type the hoppers are inactive and the individual nymph lives solitary or scattered. They are restricted in the desert breeding areas namely. Rajasthan, parts of Saurashtra and Baroda, Cutch, Hissar and Mahindargarh district in Punjab. These permanent desert homes of the locust undergoes certain changes in phase. The hoppers of solitary phase are of green colour and behave like a common grasshopper.

Gregarious Phase: In this phase, the hoppers are very active gregarious, *i.e.*, the individuals tend to remain together. They breed rapidly and form swarms which leave the breeding areas and march in group to longer distances crossing the many countries. The hoppers of this phase are generally of black colour with pink markings when young, developing black and yellow markings as they grow old. The gregarious hoppers settle into bands, which sometimes extend over several square kilometres. They move in large numbers and devastate field after field thus leaving no green vegetation which comes across their way. As there are some individuals which are intermediate in habits, therefore they have been put into a separate phase known as transient phase.

The hoppers moult 5 times and after 3-4 weeks they changes into adult. During rainy season the main breeding grounds of desert locust in India are located in and around the desert of Rajasthan. With the end of monsoon, *i.e.*, by September the desert areas get drought affected accompanied by a rise of air temperature which generally the locust avoids. Therefore, the new hoppers leave the area and are carried partly eastward by the westernly winds into Punjab, U.P., Bihar and Bengal and sometimes southward into the peninsular areas including South India. Although these flights may found till December and January and most of them are destroyed by birds. Those which survive breed in spring in Punjab and adjoining areas of U.P. The swarms that are carried westward by north-easternly winds in the direction of Pakistan and Iran breed during winter rainfall in these areas. The next generation produced in these western areas during the spring months are carried eastward into the Indian desert areas by the westernly winds prevalent in May and June. With the advent of rainfall, the summer breeding occurs in the desert and thus breeding continues year after year. They generally fly at 18-24 km per hour and can travel long distances without resting.

Adults: The description of the adults of different species is given below:

Desert Locust: The young ones are bright pink in colour but older one turn into bright yellow. They have 6-7 stripes on each eye. The legs are hard and head and legs are of yellow colour. The pink locusts are very active and are likely to cause great damage to crops.

Migratory Locust: The adults of this species come from Pakistan in summer and generally mate in Rajasthan and Gujarat. The hind wings are of yellow colour and the stripes are absent in eyes. They are now under control and may be seen in small number here and there only.

Bombay Locust: It is found in India, Cylone and Malaya. It is generally found in sporadic form and never in swarms. The hind wings are of pink colour and stripes are found in each eye. The adults of this species are found in quescent stage from October to July. In India, it is found in Mumbai, Gujarat and Tamil Nadu and only one generation is found in a year.

Control: The Indian Government has a permanent locust warning organization as part of the Central Directorate of Plant Protection, Quarantine and Storage. The staff of which is posted in the breeding areas for constant watching and receiving information regularly about locust population. The following methods may be used in controlling the swarms of locust:

1. The eggs can be destroyed by digging them. But this method is not economical.

2. The trenches may be dug at some distance in the way of marching hoppers and after falling in them may be burried alive.
3. The hoppers as well as adults may be killed by using flame throwers or with kerosene oil torches.
4. Poison bait may also be used to kill them, the same is spread around the bushes, where hoppers rest at night or on the way of matching hoppers in the day time. The following bait mixture has been found effective:

Wheat bran	—	95 parts
BHC dust	—	5 parts
Molasses	—	Little quantity

5. Dusting of 10% BHC at the rate of 20-25 kg/ha has been found effective and cheap.
6. Spraying the crop with 0.1% neem seed kernel powder was found very promising and it protects the crop for 3 weeks.

Termite

The termites are found practically all over the world and popularly known as white ants. There are two distinct categories of termites.

Subterranean: Their colony is always partly underground and starts by a pair of reproductive forms which enter the wood or earth at the time of swarming. The common Indian species belonging to this category are as follows:

(i) *Odontotermes obesus* Holmgren.

(ii) *Microtermes obesi* Holmgren.

Dry Wood Termites: Their colony is confined entirely to wood and starts by pair of reproductive forms which enter the wood above the ground at the time of swarming. The following are the important species:

(i) *Coptetennes fletcheri* Holmgren.

(ii) *Eutermes* sp.

Since *O. obesus* is the species mostly associated with the cultivated crops, hence its description is given here.

Systematic Position

Order : Isoptera

Family : Termitidae

Host Plants: This is a polyphagous pest and feeds on variety of forest trees and agricultural crops. Among the crops grown wheat, maize sorghum groundnut and sugarcane etc. are severely damaged. The other species destroy wood in its various forms, *viz.*, wooden building structure, wooden furniture and all articles containing cellulose like paper even decaying matter like cow dung.

Distribution: The pest is prevalent in all the warmer parts of the world but commonly found in the tropical regions.

Economic Importance and Damage: The food of termite includes cellulose and they eat almost every thing which contains this substance. Though it is a polyphagous pest but mainly feeds on sugarcane and wheat. Termites attack both sugarcane stalks and setts. As result of their attack, the young canes even the shoots of older canes dry up and finally the growth is adversely affected. The cane sets are eaten up from the cut ends and are rendered hollow, their eye buds are damaged resulting the poor germination of crop. In unirrigated areas the young wheat crop is damaged badly, as they destroy the roots of seedlings are finally the plants are dried off completely. They destroy the crop at night and remain hidden in their nest during day time. The attacked plants can easily be distinguished by irregular cuttings and adhering to soil particles. The fruit trees are also attacked and on branches and stem the soil channels are formed under which they live.

Life History: Three developmental stages *viz.*, egg, nymph and adult are found. They are the social insects and live in colony. There is polymorphism in adult stage and king, queen, soldiers, workers are found. The members of the colony may be divided into (i) reproductive forms which include queen, king, complimentary forms (substitute caste) and colonizing forms (macropterous forms winged), (ii) sterile forms which include the workers and soldiers.

Egg: The winged reproductives come out in swarms from the nest generally in the beginning of monsoon. Swarming usually takes place in the day time and most of the individuals of the swarms are destroyed by birds etc. The survivals mate, shed their wings and burrow in the ground to form a new colony of which they become the king and queen. Their flight is known as nuptial flight. The queen lays first batch of 10-130 eggs about a week of swarming, but later on may be five years. Eggs are small kidney-shaped and yellowish in colour. The individual egg is 1/2 mm long and hatch out in a week or so depending upon the environmental conditions.

Nymph: The newly hatched nymphs are yellowish white in colour and about 1 mm long. They eat for some time the excreta of the king and queen and later search the food. Nymphs develop into different castes in 6-13 months after 4-10 moultings.

Adults

Reproductive Forms

Queen: The queen is the largest individual ranging 6.6-8 cm in length and 1 cm in thickness. This high dimension of the queen as compared to other members is mainly due to the enlargement of the abdomen which is full of eggs. She is the mother of the colony and lives in a specially prepared royal chamber which is situated in the centre of the nest at a depth of 1-6 feet below the ground surface. The queen is wingless, creamy white in colour and her abdomen is marked with transverse dark brown bands.

King: The winged male which remains with queen after nuptial flight is known as king. There is generally a single king in each colony which is smaller than queen and remains with her in royal chamber.

Complimentary Forms: These are the apterous or brachypterous forms of both the sexes which maintain the numerical strength of colony in absence of macropterous forms. In a single colony their number may be in hundreds.

Colonizing Forms: They are generally produced in rainy season and attracted by light. They are brownish in colour with two pairs of slender, dark brown, long narrowed wings which are used for nuptial flight and after which they are shed. These forms emerge in millions from the parent colony during monsoon and after finding their mate they crawl into the soil and start the new colony. They are thus the kings and the queens of future colonies.

Sterile Forms

Workers: Among the sterile castes the workers constitute the main labour force of the colony and represent 90% of the whole colony. They are about 6-8 mm long, dirty white in colour with brown head and eyes are small or absent. Their mouth parts are fairly strong and used as the working tools for all types of odd-jobs. They work for the whole colony by storing food, looking after the young ones and building up the termitarium.

Soliders: They are slightly bigger than the workers. Their mouth parts are well developed and used for very active defence and offence, particularly the mandibles are very large, projecting and shaped for a variety of uses. Their main function is to protect the colony but also help the workers in keeping the colony neat and clean by removing the dead and sickly members. Both workers and soldiers contain sterile individuals of both the sexes.

Control

1. The removal of dead or decaying matter or dry stubbles from the field is useful because they attract the termite.
2. The use of partially decomposed manure should be avoided.

3. Irrigation water with crude oil emulsion may be used to avoid the termite attack.
4. Mixing of any one of the following insecticide before sowing in soil has been found very effective.
   (a) Endosulfan 4% dust @ 30 kg/hect.
   (b) Lindane 1.3% dust @ 25 kg/hect.
   (c) Heptachlor 3% @ dust 35 kg/hect.
5. In standing crops any one of the following insecticides may be used with irrigation water.
   (a) Chlorpyriphos 20 EC @ 3.75 lit/hect.
   (b) Gamma BHC 20 EC @ 3.75 lit/hect.
   (c) Heptachlor 20 EC @ 3.75 lit/hect.
6. Cane setts before planting can also be dipped in solution of 1% DDT or Chlordane or 0.25% BHC suspension.
7. The seed treatment of wheat with chlorophyriphos 20 EC has also been very effective, for this, use 400 ml per quintel of seed. Before application, the insecticide should be diluted with 5 litres of water and the emulsion sprayed over the seed. The seed should be turned over to ensure proper mixing and the treated seed should be left over night for drying before sowing.

Hairy Caterpillars

A number of species of hairy caterpillars occur in India which have dense tufts of hair on their body. They are the young ones of generally moderate sized moths which have often conspicuous colouration and are nocturnal in habit. A common characteristic of this group is that the moths as well as the larvae appear in large numbers during certain fixed seasons when climatic conditions are best suited to them. There are about 10 species of hairy caterpillars found in India, out of which following are the common species.

Family-Arctiidae

(i) *Amsacta moorei* Butler-Red hairy caterpillar

(ii) *Spilosoma obliqua* Walker-Bihar hairy caterpillar

(iii) *Utetheisa pulchella* Linn. Sunnhemp caterpillar

(iv) *Pericallia ricini* Fabar. Black hairy caterpillar

Family-Lymantriidae

(i) *Euproctis lunata* (Walker)

(ii) *Euproctis fraterna* (Moore) Castor hairy caterpillar

(iii) *Euproctis seintillus* (Wlk)

(iv) *Lymantria dispar*

Among this whole group of insects, the red hairy caterpillar and Bihar hairy caterpillar are the most injurious to agriculture throughout India.

Bihar Hairy Caterpillar

Scientific Name: *Diacrisia (Spilosoma) obliqua* walker

Order : Lepidoptera

Family : Arctiidae

Host Plants: It is a polyphagous insect which generally feeds on castor, til, mung, linseed, toria, mustard, some vegetable crops and medicinal plants.

Distribution: It is distributed throughout India and mostly in states of Bihar, M.P., U.P., Punjab and Rajasthan.

Nature of Damage: The damage is caused by the larvae which are provided with biting and cutting type of mouth parts. The adults are practically harmless but responsible for progeny build up. The caterpillars in early stages feed gregariously on the lower surface of leaves and skeletonize them, *i.e.*, leaving their veins only. After a week, when they are grown up disperse over the field and eat the leaves and top shoots of plants. In case of heavy infestation the crops are completely defoliated. It has been observed that crops grown in rainfed areas having light soil receive more frequent endemic and epidemic of this pest.

Life History: The pest passes through four stages, *i.e.*, egg-larva-pupa and adult in its life cycle.

Egg: The moths of hairy caterpillars appear in large numbers after the first few good showers of monsoon and are attracted to light where they can be seen and easily distinguished. These moths emerge from the hibernating-aestivating pupae in the soil. After copulation, the female moth lays eggs at night on the lower surface of leaves. They lay eggs in large prominent batches and each batch may have 50-100 eggs. Single female can lay more than 1000 eggs in its life span of three to five days. The egg mases look-like clusters of poppy seeds specially conspicuous against the green foliar back ground. The eggs hatch within 4 to 5 days.

Larva: The larvae after hatching begin to feed on the lower epidermis of leaves and skeletonize them. They remain congregated for a few days then disperse and spread over field. The freshly hatched larvae measure 5-8 mm in length and possess dense crop of hairs on their body. The feeding propensity increases with their age. It moults six times to reach maturity and measures 40-45 mm in length. It has an orange coloured body with both end black and whole body is covered with hairs. The larval period lasts for 15-20 depending upon the environmental conditions.

Pupa: The full grown caterpillar sheds its hairs and pupates in the soil inside a rough cocoon mainly composed of hairs. The pupa is exarate, brown in colour and about 20 mm in length. After a lapse of 8-15 days the moths emerge out of these pupae during active period of its life cycle.

Adult: The moth is of medium size and pale buff in colour. There are black spots on the wing and on few last segments of abdomen. The eyes and antennae are black in colour. The moth measures 25-30 mm in length with wing expanse of 45 to 50 mm. The longevity of adults range from 5-7 days.

The life cycle is completed in 29 to 43 days depending upon the prevailing conditions. There are about 5-6 generations in a year. The insect hibernates during December-February and aestivates from May to June in pupal stage. The adults emerge with the onset of monsoon and thus the life cycle is repeated.

Control: As given in red hairy caterpillar.

Red Hairy Caterpillar

Hindi Name: *Kutra, Kamla, Lal bhurli, Katra*

Scientific Name: *Amsacta moorei Butler*

Order : Lepidoptera

Family : Arctidae

Host Plants: It is a serious and polyphagous pest which generally feeds on jowar, bajra, maize, urd, mung, til, groundnut cowpea, cotton, guar and sunnhemp.

Distribution: This species is distributed throughout Northern India and the another species *A. albistriga mung* is found in southern parts of India.

Nature of Damage: The caterpillar is damaging stage which has chewing and biting type of mouth parts. Among the whole group of hairy caterpillars this is the most injurious to agriculture. The newly hatched larvae prefer to feed on the growing parts of plants gregariously and after few days they disperse in the whole field. When the vegetation of the field is finished they march out in large bands to neighbouring fields and continue to migrate from field to field. In years of severe outbreaks, fields may have to be resown in the beginning of the season and later on it becomes too late even for resowing. In July, 68 they appeared in epidemic form in eastern parts of U.P. covering more than 14,000 hectares of land. Authors experience has shown that out breaks of this are generally associated with the seedling stage of the crop. Further, their epidemics are positively co-related with the intermittent shower of rains followed by bountiful supply of food material.

Life History: Four developmental stages *viz.*, egg, **larva**, pupa and adult are found in its life cycle.

Egg: Female moth starts laying eggs after 6 to 9 hours of mating. The eggs are generally laid preferably on the under surface of leaves. They are creamy in colour and laid in prominent clusters. These egg masses look like clusters of poppy seeds specially conspicuous against the green back ground. An egg is 0.65 mm long, oval and dark brown in colour at the time of hatching. A female moth lays as many as 100-150 eggs during its life time. The incubation period (hatching time) lasts for 3-5 days and generally occurs early in the morning or late at night.

Larva: The dark coloured tiny caterpillars begin to feed immediately after hatching. They remain congregated for a few days and then disperse. Newly hatched caterpillar measures 1.5 to 2 mm long and 0.75 mm wide. It contains two black spots on dorsal side of 1st abdominal segment. The larva moults 6-7 times and becomes full grown which measures 40-50 mm in length and 6-5 mm in breadth. They are dark brown or deep orange in colour covered with reddish brown tufts of dense hairs. At the posterior and anterior ends of some larvae, a band of black hairs is also found. The larval period lasts for 15-23 days depending upon the environmental conditions.

Pupa: Generally, the full grown caterpillars begin to go down into the soil for pupation at the end of September. Before pupation a large number of larvae migrate from cultivated fields to the neighbouring uncultivated area. Here they pupate in the soil and remain there till the onset of monsoon rain in the following year. Pupation takes place in the soil in a cocoon made from larval hairs. The pupal colour is light brown which turns to dark brown after some time.

The pupa is 16-17 mm long and 7-8 mm wide. A timely rain on the eve of pupation is a pre-requisite for successful pupation for the caterpillars which require a loose wet soil for burrowing to a depth of 15 cm. The failure of rains at the critical stage will result in whole sale annihilation of the pest. The pupa remains in the soil for about 9-10 months generally from October to July. During active period of life cycle pupal period varies 7-12 days.

Adult: The adult is of medium size moth with white wings and black bands and spots on the body. The female is larger than the male. Female is 25 mm long with wing expansion of 42 mm while male is 17 mm long with wing expansion of 34 mm. The fore wings have red stripe at the margin of costal region and there are black spots on the hind wings. The moths are nocturnal in habit and positively phototrophic in nature. They survive only for 3-5 days.

The moths start laying eggs after one or two days of emergence. The life cycle is thus completed within 28-41 days. Generally, 1-2 generations are found in a year.

Control

1. Since the moths are attracted to light, the light traps should be set-up as soon as the rains start.
2. Collect the egg masses and young caterpillars in the early stage and destroy them.
3. The chemical control operation should be resorted as quickly as possible, *i.e.*, before larvae develop the migratory tendency. The following dusts have been found effective:
   (a) BHC 10% dust @ 40 kg per hectare.
   (b) Methyl parathion dust @ 25 kg/hectare.
4. The following emulsifiable concentrates are found quite effective when used at the rates given per hectare:
   (a) Endosulfan 35 EC-1.875 lit.
   (b) D D V P 100 EC-0.599 lit.
   (c) Phosphamidon 85 EC-0.500 lit.
   (d) Chlorpyriphos 20 EC-1.5 lit.
5. Spraying of neem seed kernel powder suspension (1%) makes the larvae unable to feed on the crop.

6. Thuricide (formulation of *Bacillus thuringiensis, Bert)* spray has also proved very effective in killing the larvae of this pest.
7. Following micro-organisms are the effective parasites of this insect:
   (a) *Aspergillus fumigatus Frs.* (Fungi).
   (b) Nuclear polyhedrosis and pox-like-virus.

White Grub

Hindi Name: Safed gidar

Scientific Name: *Holotrichia consanguinea Blanch. H. serrata Fabr.*

Anomala bengalensis Blanch.

Order : Coleoptera

Family : Melolonthidae

Host Plants: White grubs are polyphagous pest and feed on almost all the kharif crops like jowar, bajra, maize, groundnut, sesamum, sunflower, chillies, cotton, sugarcane, tobacco, brinjal, cucurbit and bhindi, etc.

Distribution: They are of world wide importance as noxious crop pest and cosmopolitan in occurrence. White grub has attained a status of major devasting pest of a variety of rainy season crops in India, particularly in states like Rajasthan, Gujarat, Maharashtra, Karnataka, Tamil Nadu, Bihar, U.P. and Orissa.

Nature of Damage: The damage caused by white grubs has been estimated to the tune of 40-80%. At times, the damage caused by grubs is so much so that entire stand of the crop is destroyed thereby necessitating the resowing in the field. During the last few years, white grub has posed such a serious and alarming situation in the country that it has been designated as a National pest.

The losses are inflicted by grub and the battle, the former feeds on fine rootlets, nodules in case of leguminous crops and then girdles of the main root, whereas the latter (adult) feeds on shrubs and the tress growing nearby the cultivated fields. The plants damaged by the grub gives a wilted appearance and finally dries out, while in case of beetles the attacked plants get defoliated. In case of severe grub infestation the entire crop is destroyed.

Life History: The life cycle of this pest passes through the following four stages, *viz.*, egg, grub, pupa and adult.

Egg: The beetles emerged from the soil after dusk (7.30-8.00 p.m.) followed by good premonsoon rains, which may occur from April to June. The emergence, movement and distribution of adults are mainly governed by rainfall, temperature and atmospheric humidity. Generally mating takes place, soon after the emergence of adults. After feeding and mating, the beetles get back to the soil in early morning hours and lay eggs. The eggs are generally laid singly in loose soil or in an earthen cell inside the soil upto the depth of 10 cm. The eggs are oval, creamy white when fresh and later turn to brown in colour. The mean number of eggs laid by single female ranges from 30-120. The hatching period varied from 7-14 days depending upon the prevailing conditions.

Grub: The newly hatched grub is creamy white and measures 10-12 mm in length and 2 mm in width. They feed on organic matter. There are three larval instars of the pest and the mean body length and width are 11.73 mm and 2.91 mm for the first instar, 17.88 mm and 4.73 mm for the second instar and 32.23 mm and 7.71 mm, respectively of the third instar. The full grown grub is curved C-shaped and dirty white in colour, its posterior abdomen is shiny black devoid of folds through which the intestinal contents are visible. The head is dark brown with strong mandibles and prominent thoracic legs. The larval period for the first and second instar ranges from 9.12 days and 3.4 weeks, respectively while in case of third instar it lasts for 6 weeks. Thus total larval period varies form 72 to 82 days.

Pupa: The full grown grubs move down deeper in the soil in search of moisture and for pupation. The grub constructs an earthen cell in which it passes a quiscent or prepupal stage which lasts for one to 6 weeks. The freshly formed pupa is light yellow and extremely tender, but as it grows older it turns brown and becomes some what hardened. It is of exarate type, *i.e.,* the appendages are free from the body. The pupal duration ranges from 10-27 days depending upon the type of soil.

Adult: The freshly formed beetle is cream coloured with soft white elytra with the lapse of time the colour changes to brown and elytra gets hardend. Generally the adults are lamellate and males being smaller then females. These newly emerged beetles remain within the earthen cell and do not come out of soil until the onset of summer rains.

The total life cycle is completed from 90 to 108 days and all the known species in India have one generation in a year.

Control

1. The use of light traps is the successful method for the collection of beetles during emergence in the night.
2. The repeated ploughings, preferably soon after the summer rains, help in exposing the various stages of white grubs to their natural enemies.
3. The adult beetles can successfully be controlled by treating the trees with the following insecticides:
   (a) Carbaryl 0.1%
   (b) Monocrotophos 0.05%
   (c) Chlorpyriphos 0.05%
4. The grubs may be controlled by the soil application of following insecticides:
   (a) Phoratc 10 g. @ 25 kg/hect.
   (b) Ekalux 5 g. @ 25 kg/hect.
   (c) Carbofuran 3 g. @ 35 kg/hect.
   (d) Oftanal 10 g. @ 25 kg/hect.

5. The sprinkling of chlorphyriphos 20 EC @ 5 litres per hectare offers the economic and effective control of grubs.

   The following micro-organisms are found parasisting the grubs:

   (a) Bacillus thuringiensis *(bacteria)*

   (b) Bacillus popillae *(bact)*

   (c) Beuveria bassiana *(fungus)*

   (d) Aspergillus parasiticus *(fungus)*

6. Seed treatment of groundnut kernels with chlorpyriphos @ 15-20 ml/kg of seed has been found very effective in protecting groundnut crop against white grub.

Sugarcane White Fly

Local Name: Lahi, Safed Mashi

Scientific Name: Aleurolobus barodensis Mask.

Order : Hemiptera

Sub-order : Homoptera

Family : Aleurodidae

Host Plants: Sugarcane, jowar, bajra, maize, wheat, barely etc.

Distribution: The pest is predominantly found in the state of Punjab, U.P., M.P., Bengal and Bihar.

Nature of Damage: Nymphs and the adults of this pest cause damage by draining the sap from the leaf tissues. The attacked crop looks yellowish and pinkish from a distance. In cases of severe infestation, the leaves turn black due to the deposition of fungus on the honey dew excreted by the pest. As a result of this cane tops become unsuitable even for the cattle feed.

The juice quality of attacked canes is adversely affected. There may be a loss of 30-40% in sucrose and 20-40% in total solids. In such cases it is difficult to prepare good quality gur and to get a better sugar recovery in the mills. Generally, it has been observed that ratoon crops and the low lying fields having more moisture or stagnation of water receives more infestation of this pest on sugarcane crop. Besides, in want of nitrogen the incidence of this pest is increased.

Life History: This pest exceptionally passes through four stages, *i.e.*, egg, nymph, pupa and adult to complete the life cycle. The description of these stages is as follows:

Egg: The eggs are laid on lower surface of leaves in small straight rows, parallel to mid rib having about 51 eggs in each row. A female lays eggs at different intervals in her life time and on each occasion it lays about 100 eggs. The egg is cone shaped and provided with a tiny little pedicle by which it is attached to the leaf. Each egg measures 3.3 mm in length and looks pale yellow in colour when freshly laid. Later, it changes to orange to grey dark grey and finally to black with a brownish tinge. Egg stage lasts for 7 days is summer and 13-14 days in winter.

Nymph: The nymphs, after hatching, settle down on the lower surface of leaves. They are shining yellow in colour and measure 0.31 mm in length. Finally, their colour change into black. They continuously de-sap the plant and ultimately render it pale and sickly in appearance. Each leaf may bear several hundreds of small black nymphs on its lower surface. There are three moults; before 1st moult the nymphs have 4 segmented antennae and 3 pairs of fully developed legs whereas after moulting the legs become vestigeal and antinnae disappear altogether. At this stage they are found stiking with leaves and sucking their sap. After second moult the nymph measures 2 mm in length and having vasiform orifice at its terminal end wherein the female reproductive organs are situated. The honey dew is excreted from the anus. The nymphal period lasts for 11 days in summers and 31 days in winters.

Pupa: This is the inactive stage of the pest which is covered by the white waxy powder. The pupa is flat, oval, greyish in colour and bigger in size than nymph. There is a 'T' shaped white marking on the thoracic region from where the pupa is splitted. Generally, pupal period is 7 days which prolongs in winters.

Adult: Emergence of adults usually takes place in the morning hours. They are tiny little creatures with pale yellow body and mealy wings. The male is comparatively smaller and sluggish than the female. Soon after emergence, males fly to the females for mating. Usually egg laying starts within half an hour after mating. The adults live for about 30 days.

White fly usually becomes active with the start of monsoon in July. High humidity during the monsoon period provides favourable condition for its multiplication, therefore, its maximum attack is recorded during September-October. Later, in winters its activity slackens and again becomes active in springs, *i.e.*, during March-April. Hot and dry winds which follow in May and June do not allow it to multiply till the conditions again become favourable in July. The pest completes its life cycle in 27 days during summer which, however, extends in winters.

Control

1. The crop should be properly looked after and balancely manured particularly with nitrogenous fertilizer. In low lying areas where water logging is common, ratooning of the crop should be avoided.
2. The resistant varieties of sugarcane *viz.*, Co 285, Co S-385 and 1158 should be grown in affected areas.
3. Spraying the crop with following insecticides have been found quite effective @ 1000-1200 lit/hect.

   (a) DDT 50 W.P. -5.0 kg.

   (b) Endosulfan 35 EC-1.0 lit.

   (c) Formothion 25 EC-1.0 lit.

   In cases of severe infestation two sprayings are generally recommended; one in September and the other in October.

4. **The** following insects are the natural enemies at different stages of this pest:
   (a) *Azotus delhiensis:* on pupae
   (b) *Amitus aelurolobi:* on nymphs and pupae
   (c) *Exetmcoeras delhiensis:* on nymph
   (d) *Encarsia sp.:* nymph and pupa

Sugarcane Leaf Hopper

Local Names: Alghora, Pyrilla

Scientific Name: *Pyrilla perpusilla* Walker

Order : Hemiptera

Suborder : Homoptera

Family : Fulgoridae

Host Plants: Sugarcane, wheat, jowar, bajra, barley and other plants of the graminae family.

Distribution: Generally, it is found in all the sugarcane growing areas of the country, but it is more prevalent in U.P., Maharashtra and M.P.

Nature of Damage: Both nymphs and adults suck the cell sap from under surface of the leaves and thus the attacked leaves become pale and wilted. The honey dew excreted by the pest attracts shooty mould which interferes with the proper functioning of leaves and thus renders them unfit even for the cattle feed. In a grown up crop attack of the Pyrilla causes deterioration in juice quality as a result of which sugar recovery is adversely affected and the quality and yield of gur also suffer when infestation is severe, the recovery of sugar may fall to as low as 75 per cent of the normal. In areas of severe infestation it becomes difficult to prepare good quality of gur. *Kalra* (1973) reported that the sucrose percentage in juice is reduced by 2-5 units. Ordinarily the recovery of gur should be 10 per cent but is generally less by 2.2 to 4.5 per cent from pyrilla infested crops.

The pest is active through out the year, but more damage is caused from May onwards and it becomes worst in October and November after the monsoon. It has been observed that moderate to high humidity (60-80 per cent) coupled with low maximum temperature (35 to 38) during summer months and low rainfall or long dry intervals during the monsoons are associated with the occurrence of the pest in a serious form. The ratoon fields with high level of nitrogen are generally infested more in comparison to normal fields.

Life Cycle: There are three stages in its life cycle *viz.*, egg, nymph, adult.

Egg: The female after copulation lays eggs in clusters of 2-60 generally on the underside of the leaves; in grown-up crop the eggs may be laid on the lower dry leaf sheaths also near the midrib. The eggs are covered with white cottony filaments from the anal tuft of the female and are bright green yellow or dark brown in colour, oval in shape and 1mm in diameter. The eggs hatch in 7 to 12 days during summer and 3 to 4 weeks during October-November. In winter the eggs are deposited on the leaf sheath, so that there may be no effect of cold. Generally, eggs are laid from April to November and during severe infestation they may be laid down on grasses, wild plants and even on ground.

Nymph: The nymphs which hatch out from the eggs are pale brown in colour and 1.3 mm long. At this stage they do not possess snout, wings and tufts. Within a week they are characterised by the possession of two long tufts of white waxy secretion at the end of the abdomen. They start sucking the sap of the canes, grow in size and change into adults after 5-6 instars usually, there are 5 nymphal instars. The total duration of nymphal stage may vary from 4-6 weeks in summer and 15 to 22 weeks during winter months.

Adult: The adult hopper is straw coloured measuring 9 mm long with wing expansion of 23 mm. There are two pairs of wings folded roof shaped on the head extended anterior from a beak like structure snout. The head is provided with white black large eyes and 3 segmented antennae the females can be distinguished from the males by the presence of pair of pad on the posterior end of the body. Prominent bundles of white waxy threads develop over these pads during the egg laying period. The female survives 6 to 8 weeks, while male only 4 to 6 weeks.

The life cycle from egg laying to adult stage is generally completed in 42 to 63 days and there are 4 generations in a year. The first 2 broods are clear-cut but the third and fourth overlap each other. The eggs of Ist generation are laid in first week of April, IInd generation from late June to middle of July, IIIrd generation in the middle of August and fourth generation are laid during November and December after which adults usually die due to cold. The cycle is continued through the nymphs which develop in to adults in the following March, thus completing four generations during the year.

Control

1. Burn all trash after harvesting the sugarcane.
2. Remove all the leaves bearing egg-clusters during March-April.
3. Ratooning should be avoided in badly infested areas and balance N_2 should be given.
4. The resistant varieties like Co 386 and Co 975 may be grown, if possible.
5. The following dusts have been found very effective in controlling the pest:
   (a) Malathion 5% @ 40 kg/hect.
   (b) Carbaryl 10% @ 30 kg/hect.
6. The crop may also be sprayed by any one of the following spray to protect the infestation of pyrilla:
   (a) Endosulfan 35 EC @ 1.0-1.5 lit in 1200 lit of water/hect.
   (b) Phosphamidon 100 EC@ 250 ml.

(c) Dimethoate 30 EC @ 0.750 to 1.00 lit.

(d) Malathion 50 EC @ 1.50 to 2.00 lit.

(e) Malathain 96% LVC (ULV spraying) 1 to 1.25 lit.

7. *Natural enemies:* The following are the effective natural enemies of the pest:

(a) Egg parasites:

(a) *Tetrasticus pyrillae*

(b) *Aeginiaspis pyrillae*

(c) *Cheiloneurus pyrillae*

(d) *Oonocystus pyrillae*

(b) *Egg predator:* Brumus saturalis

(c) *Nymphal parasites:*

(a) *Epipyrops melanoleuca*

(b) *Lestodryinus pyrillae*

(d) *Fungal Parasites:* Metarhizium anisopliae.

Sugarcane Top Borer

Local Name: Choti bedhak, Salai, Gabsukh and Funga.

Scientific Name: *Tryporyza (Scirpophaga)* nivella Fabr.

Order : Lepidoptera

Family : Pyralididae

Host Plants: Sugarcane, Jowar, Kans *(Saccharum spontaneum)*, Sarkanda *(Saccharum sara)* and Dhab *(Eragresits csnosureid)*.

Distribution: All sugarcane growing areas specially in U.P., M.P., Bihar, Punjab, Chennai and Mysore.

Nature of Damage: The damage is caused by the caterpillar which bears chewing and cutting type of mouth parts. Its attack starts on the growing crop from the month of May. The newly hatched larvae move on the leaf surface for 3 to 4 hr and then after selecting a suitable spot on the first open leaf on the top, penetrates in to the mid-rib from the underside of the leaf and bores its way towards the axil. It comes out to the mid-rib and bits in to whorl of newly developed leaves which cause the series of holes and may be seen when the leaves open. Thus, it goes to the spindle and moves down towards the growing point and destroy the top portion of the plant. The central shoot when tunneled and damaged by the larva, dries up and resulting in the formation of dead-heart. When the central shoot is damaged, side shoots develop and produce a sprout and branched top known as bunchy top.

Generally, it is found through out the season but major damage is caused from July to September. The loss is caused to the tune of 30 per cent. The bored cane mature early but deteriorate from January onwards when a decrease of one or two units in sucrose per cent in juice is observed.

Life cycle: Four developmental stages are found in its life cycle *viz.*, egg, larva, pupa and adult.

Egg: The female moth lays eggs in clusters of 5-8 mostly on the under side of the leaf which is covered with brown hairs from the anal tuft. Eggs are flattened, scale like and overlap one another in the cluster. A single female can lay about 500 eggs. They are round, flat and orange brown in colour measuring 1.2 mm long and 1.0 mm wide. At the time of hatching they become light brown in colour, incubation period lasts 9-12 days.

Larva: The newly hatched larva is white in colour with black head and measuring 22 mm long. After remaining on the leave it enters the top shoot of the cane through the mid-rib of the leaf grow in size and goes on penetrating the top till 4 to 6 internodes are bored. The full grown larva measures 28 mm long, 45 mm thick and creamy white to light brown in colour, It has a dirty stripe on the back. The caterpillar constructs a silken gallery for pupation and exit hole for emergence of the moth. The period taken by the caterpillar in attaining the full growth is about 3 weeks in summer and rainy season and 5 weeks in winter. It remains in hibernation during winter and stimulates activity in February.

Pupa: Full grown larva pupates inside the cane having nouth towards exit hole. The pupa is shining brownish yellow in colour measuring 16-18 mm long. The pupal stage lasts 10-12 days and out of which adult moth emerge out from exit hole.

Adult: The adult moth is silvery white, 2.5 mm long and with a wing expanse of 3 mm. The abdomen of the male is slender and pointed while that of female is stout and covered with crimson or orange hairy anal tuft. The antennae are filiform and tarsi segmented. The fore wings are long, narrow and bear hairs on the outer and posterior margin. The moths are nocturnal in habit.

The life cycle is completed from 42 to 51 days and there are 6 generations in U.P. while only five in Bihar and Punjab. The pest over-winter in caterpillar stage and regain activity in the beginning of February. The eggs of Ist generation are laid in March on ratoon sprouts or autumn planted crop, while of second generation in the month of May. The attack of IInd generation is generally mild as conditions (hot weather) prevailing at that time are not suitable for its activity. The attack of the IIIrd generation starts in July and of fourth in the middle of August. The sixth generation attack appears in the end of September or early October and caterpillars of this generation enter into hibernatrion within the cane.

Control

1. The moths may be destroyed by attracting them in light traps.
2. Adequate and timely nitrogen holds the crop to withstand the attack of top borer.
3. The egg masses may be destroyed easily as they can be spotted out due to reddish-brown hairy mass covering them.
4. A campaign may be organised to remove the top bored plants in the month of July.
5. Ratooning should be avoided where the attack of the

pest is usually serious and water-logging condition should also be avoided specially in rainy season.

6. Grow following resistant varieties, e.g., For U.P. Cos 761, 796

   For Bihar B.O. 10, 22 and 24

   The following varieties have been found better to different parts of U.P.

   Eastern-Co. 395, 453

   Western-Co. 186, 243, 421 and 443

   Central & Ruhelkhand-Co 313, 421 and 453.

7. Sprinkling of endosulfan or carbofuran granules over the top portion of the crop @ 5 and 15 kg per hectare, respectively were found quite effective.
8. Spraying the crop with 0.1% endosulfar (35 EC) @ 1200-1500 lit/hec minimises the attack of borer.
9. Recent studies have shown that soil application of carbofuran or phorate granules @ 2 kg a.i. per hect. is also very effective in controlling the pest.

Natural Enemies: The following are the important natural enemies which keep the pest population low up to certain extent:

Egg Parasite: *Trichogramrna sp.*

Larval Parasite: *Apanteles flavipes.*

Pupal Parasite: *Stenobracon deesae.*

Sugarcane Shoot Borer

Local Name: Ankur bedhak or Stambh bedhak

Scientific Name: *Chilo* (Chilotraea) *infuscatellus* Snell.

Order : Lepidoptera

Family : Pyralididae

Host Plants: Sugarcane, Maize, Jowar and Bajra etc.

Distribution: Besides India, it is found in Mayanmar, Japan and Korea. In India it is mainly distributed in U.P., M.P., Punjab, Bihar and Maharashtra.

Nature of Damage: The damage is caused by the caterpillar which bore the cane stalk at the level of the ground and feed inside the soft tissues, going upwards as well as down wards as far as the roots. The central whorl of leaves dries and forms characteristic dead-heart. The peculiarity of the dead-heart is that the only central leaf dries up and other leaves remain green for a long time. When dead heart is pulled out, it gives very offensive smell. The shoot borer caterpillar is migratory in habit and, after killing one shoot it goes to another and a single caterpillar may destroy three or four shoots. The long tunnelling of stem greatly reduces the yield and quality of the juice. Under severe infestation it may cause 25-30 per cent loss to the crop. This pest is more serious in western and central cane growing areas of the Uttar Pradesh. In Northern India, maximum damage is caused during April to June. The incidence of shoot borer is more pronounced in light soils under dry conditions.

Life cycle: Four developmental stages namely egg, larva, pupa and adult are found in its life cycle which are described below:

Egg: The eggs are laid in clusters by female moth on the under surface of the leaves. They are creamy white, oval, scale like and deposited in two or three overlapping rows. There are more than 30 eggs in each cluster. The eggs hatch in about 4 to 6 days.

Larva: Newly hatched larva possesses a dark head and translucent body with long hairs on dark spots. These spots change into prominent grey tubercle provided with long bristles and five stripes of violet colour develop. The newly hatched larvae feed singly or in groups on the tender leaf sheath for 4 to 6 days. The behaviour at this stage is more or less like a leaf miner. Later, they disperse and each caterpillar makes a hole on the stem just at the soil surface. After entering in to the cane stalk they feed on plant tissues for 3 to 4 weeks. The full grown caterpillar is 25 mm long and of dirty white in colour. Larval period lasts 3 to 4 weeks and then it changes into pupa.

Pupa: Full grown larva pupates within the tunnel in the shoot and before pupation it constructs an exit hole in the cane just above the ground surface for the emergence of moth. The pupa is brownish in colour. The pupal period lasts for a week and thereafter a moth emerges through a hole in the stem. During winter larvae do not change in to pupae but hibernate and pupate in the month of March.

Adult: The adult moth is greyish brown in colour with dark markings on the fore wings while hind wings are greyish white. Eyes large, antennae filiform with a wing expanse of 25 to 35 mm. The posterior region of the abdomen is some what pointed.

The adult moths after emergence copulate in the night and females start egg laying after 2-3 days of copulation. The life cycle is completed in 34-44 days. There are generally 5 to 6 generations in a year but during favourable conditions 8 generations in a year have been recorded. Pupation among the over wintering larvae starts in the later half of February or beginning of March when there is slight rise in temperature.

Control

1. The stubbles may be destroyed and burnt after harvest.
2. Earthing of sugarcane should be done during May and June and the surface portion of cane may be covered.
3. Timely cutting of attacked shoots at or just below the ground level ensures destruction of most of the caterpillars present within the stem.
4. Larvae may also be destroyed by pulling the dead hearts.
5. Early planting (by middle of February) is helpful in checking serious damage of shoot-borer under North Indian conditions.

6. Treat the soil with 10% BHC dust @ 30 kg/hec before sowing.
7. Sprinkling of gamma BHC, Heptachlor or Telodrin emulsion at 1 kg a.i. per hec mixed with about 1,500 litres of water over the sets in the furrows at the time of planting of cane proved effective.
8. The spraying of endosuflan 35 E.C. @ 1.85 to 2.50 lit in 1200 lit of water was found effective. This should be repeated 3 to 4 times at 10-12 days intervals.

Natural Enemies: The following natural enemies have been found on different stages of the pest:

Egg Parasite

(a) *Phanurus beneficiens*

(b) *Trichogramma minutum*

(c) *T. nana*

Larval Parasite

(a) *Apanteles flavipes*

(b) *Stenobrancon nicevillei*

(c) *Goniojus indicus*

Pupal Parasite

(a) *Tetrasticus ayyeri*

(b) *Aperostocus sp.*

(c) *Melcha ronetipinis*

Gurdaspur Borer

Local Name: Gurdaspur, bedhak, Dehradun bedhak

Scientific Name: *Acigona (Bissetia) Steniellus Hmpsn.*

Order : Lepidoptera

Family : Pyralididae

Host Plants: Sugarcane, Jowar and Maize etc.

Distribution: The pest was originally recorded from Sialkot (Pakistan) and Gurdaspur district (Punjab) in the early twenties.

Now it is found in all sugarcane growing areas specially in Punjab, Haryana, Uttar Pradesh, Rajasthan and Madhya Pradesh. In U.P. it's incidence was found in Saharanpur, Muzaffarnagar, Dehradun and Rampur districts.

Nature of Damage: The attack of Gurdaspur borer starts with the commencement of monsoon rains in the later part of June or July. Damage is done by the caterpillar which is quite characteristic. The caterpillar after entering the internode, moves upwards in a spiral manner making minute holes on the bud from within the internode, with the result that canes become weak, dry up or fall down easily in strong wind or heavy rain. Later on it bores in to central portion of the cane and feeds in a single straight tunnel moving upwards.

At the beginning larvae feed gregariously but after 10 to 12 days they disperse to adjoining canes. As the attack proceeds, whorl of leaves at the top dries up due to which further growth ceases. The damage normally varies from 15-25 per cent of the total crop, but in years of severe attack it may be 40 to 50 per cent in certain fields. *Kapoor* (1957) estimated the loss caused by this pest up to the extent of 38-100 per cent.

Life Cycle: There are four developmental stages of the pest namely egg, larva, pupa and adult.

Egg: The eggs are laid by female in clusters in and along the groove of the midrib on the upper (ventral) surface of the leaf near the leaf sheath. Newly opened leaves in the top portion of the shoots, are preferred for egg laying; usually each cluster contains 20 to 50 eggs. Initially they are creamy white in colour which subsequently change in to brown. The eggs are flat and 1 mm in diameter. The egg stage lasts for 2-5 days.

Larva: The newly hatched larva after hatching feeds for some time on the bud situated on the top first or second internode and then bores into it from near the bud. During this stage as many as 40 to 50 larvae may be found in an internode. The larvae are pinkish in colour measuring 2.5 mm long. They come out after a period of 7 to 10 days and disperse to surrounding canes, each caterpillar penetrates separate cane. The full grown larva is 30 to 35 mm long having 4 violet stripes on the back and pinkish brown in colour. The larval period is of 21 to 34 days.

Pupa: Pupation occurs inside the cane. The pupa is brown in colour measuring 10 mm long and 35 mm wide. The pupal period lasts 6-7 days after which moth comes out through the exit hole prepared by the caterpillar.

Adult: The moth is greyish-brown with dark markings on the fore wings and measuring 10 mm long and 3.5 mm wide. The hind wings are greyish white with wing expanse of 25 mm.

The moths copulate on 2nd day of emergence and female starts egg laying after 2 days. The total life cycle is completed in 31 to 48 days and there are 3 generations in a year. The pest remains active from July to November. From July to October all stages of the pest are seen simultaneously because of overlapping generations. After November the caterpillars remain in hibernation during winter and in quiscent stage during early summer within the sugarcane stubble and become active after the first shower of rain.

Control

1. The attacked shoots should be regularly removed and destroyed.
2. The rougeing should be done from July to September, when young caterpillars feed gregariously.
3. Resistant varieties should be preferred, *e.g.*, Co 1157, 1158 and Co 6812.
4. In endemic pocket, the setts of cane should be treated for 2 hours in 0.2% trichlorfon (dipterex) before planting.
5. The following insecticides have been found effective in minimising the incidence of pest in standing crop. These may be sprayed 4 times at 15 days intervals @ 1200 to 1500 litre of water:

(a) Gamma B.H.C.3.75 lit.

(b) Endosulfan 35 EC-2.0 to 2.5 lit.

Natural Enemies: The following are the effective enemies in different stages of the pest.

Egg Parasite: *Trichogramma sp.*

Larval Parasite

(a) *Apanteles flavipes*

(b) *Struniopsis interens*

Pupal Parasite: *Isotima javensis*

Sugarcane Root Borer

Local Name: Mool bedhak, Eekh ke Jar Ki Surhi, Jarkata, Makora.

Scientific Name: *Emmalocera depressella* Swinh.

Order : Lepidoptera

Family : Pyralididae

Host Plants: Sugarcane, Maize.

Distribution: It is found in all sugarcane growing areas of the country specially in U.P., M.P., Punjab, Bihar and South India.

Nature of Damage: The damage is caused by the caterpillar during early stage of the crop. Its damage starts during April and caterpillars of the pest bore at the base of the newly formed stem and feed inside. As a result of which main shoot is damaged and dead-heart is formed. The damage is very peculiar, because larvae enter at the base of the stem which is very close to the surface of the soil. Further, they do not move upwards but remain near the base of the shoot. The dead heart caused by the root borer resembles that of shoot borer but may be differentiated on the following points:

1. Only one hole is found on the shoot.
2. Absence of offensive smell in dead heart.

3. Alongwith central leaf sheath, few side leaves are also dried and involved in dead heart.
4. There is no sign of attack (eating holes) on the leaf sheath.

The attack of the pest is more pronounced in ratoon crops and it persists from April to the end of September or middle of October. If the plants are young they may die but the older ones dry up and fall down. During rainy season its attack is minimised. Due to the attack of this pest there may be a decrease in yield by about 10 per cent reduction in sucrose percentage in juice by about 0.3 unit. In cases of severe infestation it may cause a loss to the tune of 40 per cent.

Life Cycle: Four developmental stages namely egg, larva, pupa and adult are found in its life cycle, which are described as under:

Egg: The female lays eggs after copulation in the night on the underside of the leaves of the sprouting canes during April and May and lower most part and even on the soil surface. The eggs are deposited either singly or a few of them in clusters. They are flat, oval, creamy white in colour and measuring 1.2 mm long and 1 mm wide. The hatching period is about 7 days.

Larva: The newly hatched larva is light yellow or white in colour with a brown head measuring about 2.00 mm long. The larva enters the plant near the roots by making a small hole and feeds inside, if plant dries out it migrates to the another plant. It moults 4 times and becomes full grown which is creamy-yellow in colour with brown head and measuring 30-35 mm long. The larval period lasts for 40-45 days and before changing in to pupa it constructs a silken tube for emergence.

Pupa: The pupa is reddish-brown in colour measuring 17-20 mm long and 4 to 5 mm wide. The pupal stage is not found during winter. Generally, the pupal period lasts in 7-10 days which may extend up to 15 days depending on the environmental conditions.

Adult: The adult moth is dirty brown in colour, measuring 17 to 20 mm long with the wing expanse of 27 mm. The hind wings are white, shorter and wider than the fore wings and have a dark longitudinal stripe on each wing. The head is small with large black eyes and bipectinate antennae in the case of male and setaceous in case of female.

The female starts egg laying after 2 days of emergence. The total life cycle is completed in 57 to 73 days and there are 4 generations in a year, however 5 generations have been found in Punjab. In U.P. the larvae of 4th generation remain in hibernation during winter. The over wintering larvae are revived to activity in April when moths of 1st brood emerge and start egg laying. The maximum activity has been found in the month of August.

Control

1. Timely cutting of infested shoots at or just below the ground level is found useful
2. After harvesting stubbles should be destroyed and ratooning be avoided in endemic areas.
3. The resistant varieties *viz.*, Co 313, 513, 1148 and 1158 should be preferred to grow.
4. Sprinkling of gamma BHC @ 6.05 lit per hectare mixed with about 1875 litres of water over the sets in furrows at the time of planting found effective.
5. At the time of planting dusting of 10% BHC dust @30-35 kg per hectare in furrows was also found effective.
6. The spraying of endosulfan @ 1.85 to 2.50 litres/hectare in 1200-1500 lit of water gave good control when applied 4 times in standing crop at an interval of 10-15 days.

Natural Enemies: The following are the effective natural enemies of the pest:

Egg Parasite

(a) *Trichogramma minutum*

(b) *Phanurus beneficiens*

Larval Parasite

(a) *Stenobracon deesae*

(b) *Phanerotoma sp.*

(c) *Chelonus narayani*

(d) *Apanteles sp.*

Jowar and Maize Stem Borer

Local Name: Mokora, Makaya, Obra, Ratwa, Kanka ghundi, Khera, Tana chhedak.

Scientific Name: *Chilo partellus (zonellus)* Swinhoe

Order : Lepidoptera

Family : Pyralididae

Host Plants: Jowar, Bajra, Maize and Sugarcane etc.

Distribution: It is found practically in all the tracts of India where these crops are grown. Its distribution extends in the region of the world from Afghanistan to Indonesia and Taiwan in the east and Sri Lanka in the south and it has also been reported from east Africa.

Nature of Damage: The freshly hatched larvae begin to feed on the tender leaves, particularly in the central whorl; after feeding for few hours on leave, they enter through the mid whorl into the shoot. The caterpillar attacks all parts of the plant except the roots. In case of maize only young plants while in jowar all stages of plants are attacked. In case of younger plants the growing point and the base of the central whorl are infested resulting in the drying up of the central whorl and forming the dead heart. In case of older plants, the stem is riddled with.

The attacked plants remain stunted in growth and produce no grain. More than one caterpillars are found in a plant. Generally 20 to 30 percent damage is caused by the pest and it is serious in dry years. *Jotwani* (1971) estimated 83.7 per cent loss in grain yield. Maximum damage is caused in the month of August to maize while during September and October to jowar. The pest remains active in the field from March to November.

Life Cycle: Four developmental stages *viz.*, egg, larva, pupa and adult are found in its life cycle.

Egg: The female moth starts egg laying after two days of copulation. The eggs are laid in clusters on the lower surface of the leaf near mid-rib which are scale like and yellowish creamy white in colour. Each cluster may have about 4 to 20 eggs and overlapping one another in disposition. An individual egg is 1 mm long and turns in to brown and black at the time of hatching. The hatching period varies from 2 to 5 days. A female lays 250-300 eggs in her life span.

Larva: The freshly hatched larva is white in colour and about 2 mm in length. The head and prothorax are black and body is provided with four long brown stripes on its back. They feed about a week on the leaves then enter in to the stem and remain inside whole life. Generally, larval period is of 21-28 days but it extends to 193 days in winter as the caterpillar remains in hibernation. The full grown larva is dirty white in colour and having black spots on the body which are provided with a brown hair. It measures 27 mm in length.

Pupa: Pupation takes place in specially constructed chamber in the stem. Pupa is reddish-brown in colour and the tip of its abdomen is furnished with six spines arranged in a group of three. The pupa is about 10 to 13 mm in length. The male pupa is narrow and smaller than that of female. The pupal period lasts in 4-7 days and after which the moth emerges from the stem through a hole made by the larva at the time of pupation.

Adult: The adult moth is medium sized insect with a wing expanse of 25 to 30 mm and yellowish brown in colour. The fore-wings are pale straw coloured while the hind-wings possess a double row of black spots along their outer margins. The tip of the abdomen in female is dilated and tufted with hairs, while in male it is pointed and devoid of tuft.

The adult moths copulate after emergence. The male dies after copulation while female after 2 to 3 days of egg laying. They are nocturnal in habit and during the day remain hidden under dried leaves, clods of earth etc. The life cycle is completed in 29 to 41 days and there are 4-5 generations in a year. The moths from over-wintering larvae appear in March-April and start egg laying.

Control

1. The stubbles should be collected and destroyed, when harvesting is over.
2. Collection and destruction of all the dead hearts and attracting the moth in light-trap proved effective to control the pest.
3. The following insecticides have been found effective when sprayed in standing crop at the interval of 2, 5 and 7 weeks of sowing.
   (a) Dimecron 100 EC-200, 250 and 300 ml/hect.
   (b) Endosulfan 35 EC-1.0, 1.25 and 1.5 lit/hect.
   (c) Carbaryl 50 W P-1.0, 1.5 and 2.0 kg/hect.
4. The placing of following granules in the central whorl of the plant has been found quite promising but this application should be followed by second granular application or spraying with any one the above insecticides after 3 weeks of placement:
   (a) Lindane 6% granule-20 kg/hect.
   (b) Endosulfan 4% granule-15 kg/hect.

Natural Enemies: The following are the effective natural enemies:

Egg Parasite

(a) *Trichogramma minutum*

(b) *Bracon chinensis*

Larval Parasite

(a) *Microbracon chinensis*

(b) *Apanteles flavipes*

(c) *Xanthopimpla sp.*

(d) *Blepharipoda sp.*

Sorghum Shoot Fly

Scientific Name: *Atherigona socata* Rond

Order : Diptera

Family : Anthomyidae

Host Plants: Sorghum, maize, wheat, broom corn, small millets and grasses.

Distribution: It is widely distributed in Europe, Africa and Asia. In India the pest is more serious in Southern parts.

Nature of Damage: The maggots cause damage to the seedlings as well as to the early stage of the crop when it is in the six leaf stage. After six weeks of sowing the crop is rarely attacked. The maggots feed on the main shoot with the result growing-point is destroyed. The death of central short causing a typical symptom called "dead heart". The damaged seedlings may be killed or it may produce few non-productive tillers. When the plants are attacked in some what older age, the tillers are produced which mature later than the main crop. It may cause loss to the crop as high as 60 per cent. Extremes of temperature as well as heavy continuous rains adversely affect the population of shoot fly.

Life History: The four developmental stages namely egg, larva, pupa and adult are found which have been described below.

Egg: The female lays whitish cigar-shaped eggs, measuring about 1.2 mm in length, singly on the underside of the leaves of seedlings. Generally, only one or two eggs are laid on each seedling. The fly prefers to lay eggs on the central leaves. The eggs are provided with two wing-like lateral projections. The incubation period of eggs varies from 1-2 days.

Larva: The newly hatched tiny maggots creep out and reach in between the sheath and the axis, and bore into the shoot. The maggot is dirty whitish in colour and develops inside the shoot by feeding on decaying tissues. The larval period lasts for 6-10 days. The full grown larva is yellow coloured and measures 5.5 to 5.6 mm in length.

Pupa: The full grown maggots pupate either inside the stem or come out and pupate in the soil. The pupal period lasts for 6-9 days. The colour of pupa changes from initial light brown to deep brown. Pupa is of object type and its anterior end is round and is fitted with circular disc bearing the two spiracles.

Adult: The adult shoot fly is about 5 mm in length and be identified by the presence of three pairs of black spots on pale grey abdomen. They are more active in the early mornings or during evenings. The adult usually survive 3-4 days.

The life cycle of fly is completed in 14-23 days and many generations are found in a year. In northern India, the pest over-winters in the pupal stage.

Control

1. High seed rate may be used.
2. Roguing of attached plants on the appearance of dead hearts.
3. The resistant varieties *viz.;* CSH-7 and CSH-8 may be grown to reduce the damage.
4. Application of phorate or disulfoton granules in the furrows at the five on sowing @ 1.5 kg a.i./hect.
5. Seed treatment with carbofuran at the rate of 5 parts insecticide in 100 parts of seed is also quite effective.
6. Sprayings of 0.025% methyl demeton or phosphamidon is also effective, this should be repeated after 3 days intervals.

Earhead Caterpillar

Scientific Name: *Cryptoblabes gnidiella* Mab

Order : Lepidoptera

Family : Pyralididae

Host Plants: Jowar, Ragi and Bajra.

Distribution: The pest is distributed through out the country, where sorghum is grown.

Nature of Damage: The damage is caused by larvae which feed initially on lemina by scraping the chlorophyll and lateron on the milky as well as hard grains of sorghum. The caterpillars web together adjacent grams in a silken thread and feed on the grains inside the earheads. Heavy infested grains found to have more webs and excreta of the pest. The damage is caused generally from September to November.

Life History: The four developmental stages, *viz.*, egg, larva, pupa and adult are found in its life cycle which have been described below.

Egg: The female lays eggs on spikelets and on grains of the earheads. Freshly laid eggs are creamy unite in colour, round or conical, flattened in shape. The individual egg measures 0.45 mm in length and 0.32 mm in breadth. They are laid singly. The incubation period varies from 3-4 days.

Larva: The newly hatched caterpillar is dirty white in colour with brown head. Body is translucent without strips consisting of 12 segments. They feed on the developing grains and grow in size. The full grown larva is active measuring about 12 mm in length and 2 mm in breadth with dark brown colour. The larvae form web come out of the excreta and the silken thread on the cob and remain inside. The larval period lasts for 9-10 days.

Pupa: Pupation takes place in the silken webs. The pupa is short and brown in colour, covered with silken membranous webbing threads which are secreted by the caterpillar. The pupa measures about 7.5 mm in length and 1.8 mm in breadth. The pupal period lasts for 7.8 days.

Adult: The adult measures 6.5 mm in length with using expanse of 7 mm. The thorax is dark grey in colour. The hind wings are having fringe like hairs on anterior margin and are transparent. The fore wings are smaller than hind wings and dark-grey in colour. The longevity of male and female moth in 3 and 5-6 days, respectively.

The newly emerged adults mate in the morning hours and start egg laying after 20-24 hours of mating. The total life cycle is completed in 20-22 days and there are 4-5 generations in a year.

Control: Spraying the crop with any one of the following insecticides:

(i) Endosulfan — 0.07%
(ii) Monocrotophos — 0.05%
(iii) Phosphamidon — 0.03%

3

Major Useful Elements

Abscisic Acid

Abscisic acid was discovered as a powerful inhibitor (inhibitor P) of seed and bud germination. It was known to be of widespread distribution in plant kingdom, but it was isolated for the first time in crystalline form from mature cotton fruits by W.C. Liu and H.R. Cairns in 1957 only. The chemical structure of the compound, was however, not determined at that time. Since the compound stimulated abscission of cotton petioles, it was called *abscisin* or *abscisin I.* Subsequently *abscisin II* was isolated from young cotton fruits by K. Ohkuma and co-workers in 1963. Almost at the same time C.F. Eagles and P.F. Wareing (1963) extracted an inhibitor of apical growth from birch *(Acer pseudoplantanus)* leaves, which they called *dormin.* Later on, studies revealed that all these compounds had almost similar chemical nature.

Partial chemical characterization of abscisin II revealed that it was a 15-carbon compound. J.W. Conforth (1965) and his associates demonstrated that abscisin II and dormin were the same compound: the S-abscisic acid (AbA). Abscisin I was also known to be the same compound, later on. The principal component of inhibitor β described by T.A. Bennet-Clark and N.P. Kefford (1953) was also AbA.

The name abscisic acid was given to the compound because of its known effect on stimulating abscision. However, it is now well established that AbA is much more important in causing dormancy of both buds and seeds and in stomatal control than in abscision.

Structure and Distribution

Abscisic acid is a sesquiterpene consisting of 15-C atoms. It is unique among plant hormones in having an asymmetric carbon atom. It has a six carbon ring structure to which a side chain is attached. Because of the asymmetric carbon atom (carbon-1), it occurs in two enantiomorphic forms, R-abscisic acid and S-abscisic acid. The naturally occurring form is S-abscisic acid. This form of AbA is dextro-rotatory (+). The degree of optical rotation of + AbA is perhaps the highest among known biological molecules. Because of this property, AbA is often detected by optical rotation dispersion technique. The AbA has been synthesized in laboratory also and synthesized AbA is generally a racemic (±) mixture.

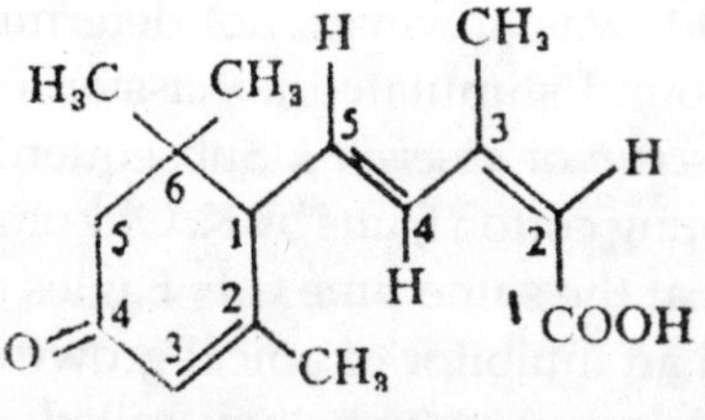

Abscisic acid molecule

AbA is common in plant materials, ranging in concentration between 0.03 to 4.0 mg/kg fresh weight. On molar basis the concentration ranges between 0. 001 and 1 μM. It is of general occurrence in monocots, dicots, gymnosperms and some ferns. Generally fruits and seeds contain the highest amount of AbA. However, it is easily transportable to other tissues as well. In fruit itself, it comes from its site of synthesis, the mature leaves. Most likely the translocation takes place in the phloem and xylem.

AbA has been isolated from ferns, gymnosperms and angiosperms, where it accumulates in appreciable quantities. It has also been found that the fungus *Cercospora rosicola* synthesizes and secretes large amounts of AbA in the culture medium. AbA is apparently not present in liverworts. But another compound—lunularic acid seems to play a role similar to AbA in liverworts.

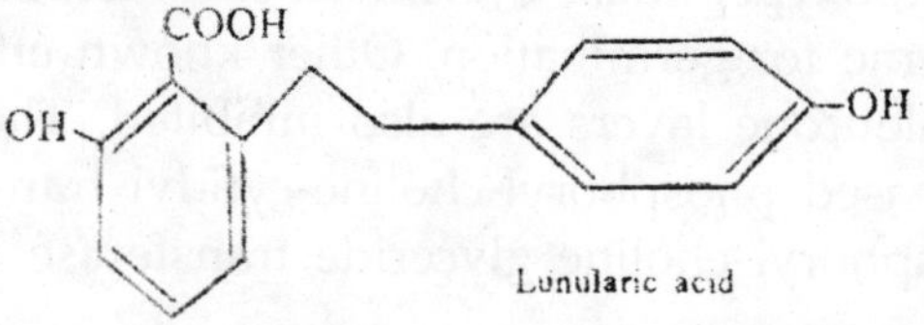

COOH
OH
OH

Lunularic acid

Bioassay: Many bioassays have been developed for detecting AbA. Inhibition of seed germination is often used to detect the presence of AbA in plant tissues.

Another bioassay is the test on elongation growth of seedlings. In some cases, about 1 mM AbA can cause as high as 50% inhibition of the elongation growth in 2 h.

Other bioassays of AbA include inhibition of a-amylase synthesis, stomatal closing in epidermal peels, and acceleration of abscission in excised tissues. For the test of alpha amylase inhibition, the excised barley aleurone layers are incubated in buffer containing GA for 3-4 hours. GA causes a substantial increase in the enzyme activity. However, when AbA or the test solution suspected to contain AbA is also included in the incubation medium, the GA mediated increase in enzyme activity is abolished. A similar response is also seen with cycloheximide, an inhibitor of protein synthesis.

Physiological Effects

Inhibition of Seed Germination: Exogenous supply of AbA inhibits germination of most non-dormant seeds, for causing an inhibition, AbA must be continuously present. As soon as it is removed by washing of the seeds, the germination can take place. Endogenous AbA also inhibits germination. The AbA content of dormant seeds is high and it decreases during germination. Several possibilities have been suggested for the inhibition of seed germination:

(a) The inhibition is possibly because of inhibition of enzymes involved in germination process. The hormone inhibits protease and amylase and GA induced amylase in barley aleurone layers. It also inhibits the synthesis of carboxypeptidase Q, which is considered as a specific enzyme for germination. Other known effects of GA on aleurone layers are also inhibited. These include increased phosphoryl-choline-cytidyl transferase and phosphoryl choline glyceride transferase activities.

(b) The hormone also appears to be inhibiting water uptake by germinating mustard seeds.

(c) M. Coccuci and N. Negrini (1989) from their studies with radish seed germination have concluded that "AbA inhibits seed germination by inhibiting some activity depending on calcium-calmodulin system."

Inhibition of Seedling Growth: AbA inhibits growth of the seedlings in some cases. In *Glycine max* for example, 1 mM AbA inhibits seedling growth by about 50% in 48 h. In some cases, similar growth inhibition is observed in almost 2 h. Auxin stimulated elongation growth in coleoptile is also inhibited by AbA. It has been suggested that AbA inhibits seedling growth by decreasing the water potential.

Inhibition of Bud Growth: Exogenous AbA induces but dormancy in woody plants. As in dormant seeds, AbA content in dormant buds is high and it decreases by treatments which lead to the breaking of dormancy. The hormone inhibits lateral growth of the bud, as has been reported in tomato. Plants in which bud growth is inhibited by far-red treatment, also have high AbA content. Further, cultivars with

high apical dominance, contain high AbA levels. It appears that this high level of AbA is maintained by the IAA content of the tissue, as IAA has been invariably implicated in apical dominance.

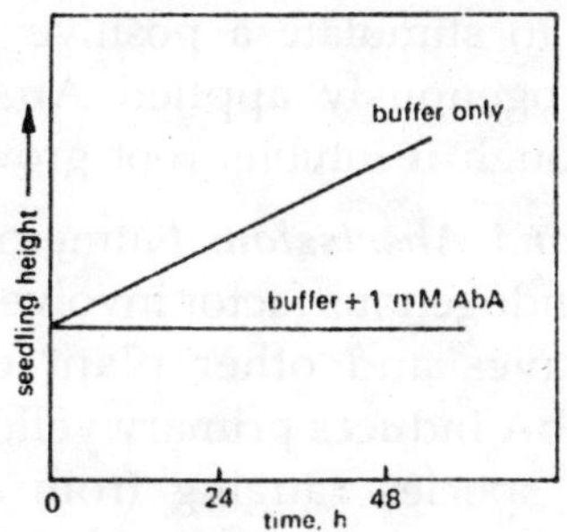

A typical response of abscisic acid on seedling height in soyabean.

Stomatal Closing: The most significant and best known effect of AbA is its control of stomatal closing. Exogenous application of AbA to epidermal strips causes stomatal closure. In fact, factors such as water stress, chilling, etc. which induce stomatal closure, also increase AbA content of guard cells. The response of AbA on stomatal closing is very fast and occurs within a few minutes of AbA application.

Regulation of stomatal opening by AbA involves K^+, H^+ and Cl^- exchange. These ions are involved in stomatal opening. AbA inhibits K^+ uptake into guard cells and has been shown to inhibit both K^+ uptake and proton release in a variety of tissues. The fungal toxin fusicoccin (FC) overcomes the effect of AbA on stomata, as well as other tissues, apparently by stimulating the H^+/K^+ exchange. Whether the AbA inhibits the H^+/K^+ exchange directly or through some other molecule is not known.

AbA also increases malate leakage from epidermal strips. Malate may be the primary source of protons necessary for the H^+/K^+ exchange process during stomatal opening. Release of malate from the guard cells, which may occur together with K^+, will induce stomatal closure not only because of reduced turgidity but also because of loss of the source of protons.

Geotropism: There seems to be some involvement of AbA and perhaps other inhibitors as well in geotropism. Appreciable amounts of AbA have been detected in maize root tips. The accumulation of AbA in the tip appears to require light and gravity. AbA produced in the cap seems to be translocated basipetally and to stimulate a positive geotropic response. Further, the exogenously applied AbA induces positive geotropism although it inhibits root growth.

Senescence and Abscission: Numerous studies indicate that AbA is an endogenous factor involved in senescence and abscision of leaves and other plant organs. Exogenous application of AbA induces primary yellowing in leaf tissues in a variety of species ranging from deciduous trees to herbaceous plants such as *Tropaeolum.* It also accelerates abscission in leaves in a number of species. In many studies, it has been found that a non-volatile substance is released during senescence of leaves, which accelerates abscission. D.J. Osborne has called this substance as *senescence factor* (S.F.). Many scientists have suggested that this senescence factor is AbA.

Exogenous application of AbA promotes loss of chlorophyll, protein and RNA and promotes symptoms of senescence in detached leaves, although relatively higher concentrations (1 to 5 mM) of AbA are required. Further, AbA content increases during senescence of detached leaves. For example, in detached tobacco leaves the AbA content increases by about 6 folds in light and 22 folds in dark in 7 days. Further, the application of kinetin which inhibits senescence prevents the rise in AbA content. However, the effects of AbA on attached leaves are not clear cut.

Role of Abscisic Acid in Drought Resistance: AbA seems to play an important role in plants during water stress and during drought conditions. Following observations suggest such a role:

(1) AbA promotes stomatal closing and thus reduces water loss due to transpiration.

(2) AbA content of mesophytic plants increases when the plant is subjected to water stress and the content decreases when the plant is no longer under stress.

(3) In many cases physiological effects of water deficit are similar to those of AbA treatments.

Some attempts have been made to correlate the ability to produce AbA with drought resistance. In some cases (but not in all), more AbA is found to accumulate in drought resistant cultivars than in sensitive cultivars. However, it will be more reasonable to say that AbA is involved in drought resistance at least in a significant proportion and other mechanism(s) of drought resistance is also operative. D.C. Walton (1980) has postulated that establishment of varieties that produce high levels of AbA may be highly significant to the growing of the crops in arid parts of the world.

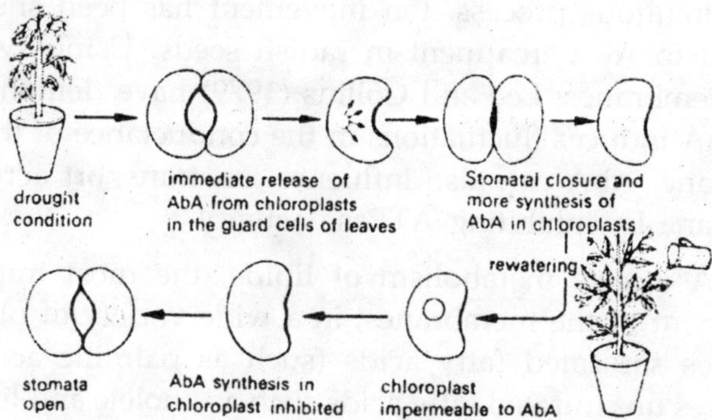

Regarding the possible mechanism of increased AbA content in water stressed plants, there are two clear cut options : (i) AbA is synthesized *de novo* from its precursors during water stress (ii) More Aba is released from the chloroplasts during the stress. Both options seem to be interrelated. While immediate responses (such as stomatal closure) may be because of the movement of AbA from chloroplasts to the guard cells, the long term responses seem to be due to more synthesis of AbA during stress. AbA is synthesized in the chloroplasts and its loss from the chloroplasts during stress promotes additional hormone synthesis. This promotion continues until the water stress is relieved off. When the water stress is relieved, the chloroplast membranes become impermeable to AbA and AbA is restricted inside the chloroplasts. Presumably further synthesis of AbA inside the chloroplast is turned off by end product inhibition.

Mechanism of Action

Membrane Permeability: Some physiological effects of AbA may be related to its effects on structure and functions of the membranes. The fast effects of AbA are apparently mediated via changes in membrane permeability. The auxin mediated elongation growth in coleoptiles can be inhibited by AbA within a few minutes. The stomatal closing is also observed within a minute of AbA application. Evidently such rapid effects are independent of nucleic acid directed protein synthesis or some other circuitous process. Ion movement has been shown to respond to AbA treatment in radish seeds. Using synthetic lipid membranes, Lee and Collins (1979) have demonstrated that AbA induces fluctuations in the conductance of the lipid membrane. AbA can also influence ion transport across the membrane by inhibiting ATPase activity.

AbA affects metabolism of lipids, the most important component of the membranes, in a wide variety of plants. It increases saturated fatty acids (such as palmitic acid) and decreases unsaturated fatty acids such as linoleic and linolenic acids. This may change the permeability of membrane.

Nucleic Acid Directed Protein Synthesis: Perhaps the long term effects of AbA on growth involve suppression of gene expression and protein synthesis. This may be achieved either by (i) decreasing m-RNA level (transcription process) or by (ii) inhibiting the translation process.

Decrease in m-RNA may be brought about either by decreased RNA polymerase activity or by increased ribonuclease level. Increased ribonuclease may be factor in the induction of senescence by AbA in some cases. Decrease in total RNA content by AbA treatment has also been observed in some cases. In maize coleoptiles, inhibition of RNA synthesis can be detected after 3 hours of AbA treatment. In bean seedlings, a 2 hour treatment with AbA decreases RNA synthesis by about 80%.

AbA may inhibit the formation of certain enzymes without inhibiting the total protein synthesis. Inhibition of seed germination and seedling growth by AbA can be attributed to the specific inhibition of some enzymes involved in the process, by the hormone. J.N. Ihle and L. Dure III (1972) demonstrated that AbA inhibited production of protease and isocitratase during precocious germination of cotton seeds. This was due to inhibition of translation of pre-existing m-RNA. AbA also inhibits the formation of α -amylase in barley aleurone layers. This is again because AbA inhibits the translation of amylase m-RNA. The formation of glutamate dehydrogenase is also inhibited by AbA. Kinetin reverses this effect.

The inhibition of protein synthesis does not appear to be an universal phenomenon. The synthesis of storage proteins in *Brassica napus* embryos is totally dependent upon the presence of AbA (Crouch, Tenbarge, Simon and Ferl, 1983). When immature embryos which have the potentiality of synthesizing storage proteins are incubated in the presence of AbA, an increase in storage proteins and their corresponding m-RNA is observed.

Involvement of a Secondary Messenger: Evidences have been put forward for the involvement of a second molecule such as calcium and calmodulin in the mechanism of action of abscisic acid. Involvement of Ca^{2+} in AbA stimulated geotropism in maize roots has been observed. M. Coccuci and N. Negrini (1989) from their studies with radish seed germination have concluded that AbA inhibits germination by controlling the development of some activity dependent on calcium-calmodulin system. In stomatal closing by AbA also, the Ca^{2+} is involved. In the absence of Ca^{2+}, AbA is unable to close stomata.

Abscisic Acid Metabolism

Biosynthesis: Two pathways for the biosynthesis of AbA have been identified: (i) From mevalonic acid and (ii) From the oxidation of carotenoids.

Direct synthesis of AbA from mevalonic acid through farnesyl pyrophosphate has been demonstrated in many cases, specially in the water stressed tissues. The water stress increases AbA formation. The early reactions in the synthesis of AbA are identical to those of gibberellins, sterols, carotenoids and other isoprenoid compounds:

Mevalonic acid → Isopentenyl pyrophosphate → Farnesyl pyrophosphate → Abscisic acid

Indirect synthesis of AbA takes place by the oxidation of carotenoids or oxidized carotenoids (xanthophyll). The xanthophyll, vialaxanthin has been shown to be the precursor of AbA in bean tissues. Oxidative cleavage of this xanthophyll either by a photooxidation process or by the action of lipoxygenase, produces a C-15 aldehyde xanthoxin. This aldehyde is then oxidised to produce AbA:

Vialaxanthin —Lipoxygenase→ Xanthoxin —Dehydrogenase→ AbA

Y. Li and D.C. Walton (1990) have demonstrated that in water stressed dark grown leaves of bean, vialaxanthin is converted to AbA by several possible pathways.

Abscisic acid was discovered as a powerful inhibitor stems and green fruits are important sites of AbA formation. It may also be synthesized in proplastids, or plastids present in the endosperm and roots.

Inactivation and Catabolism of Abscisic Acid: Exogenously applied ^{14}C-AbA is rapidly conjugated in plants to form glucose esters of AbA. This ester seems to be quite stable in plants, and possesses hormonal activity similar to AbA.

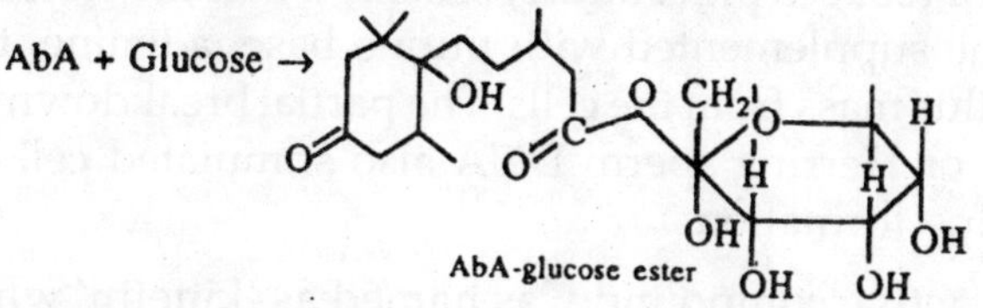

AbA + Glucose →

AbA-glucose ester

Aba is metabolised to phaseic acid and dihydrophaseic acid in some plants;

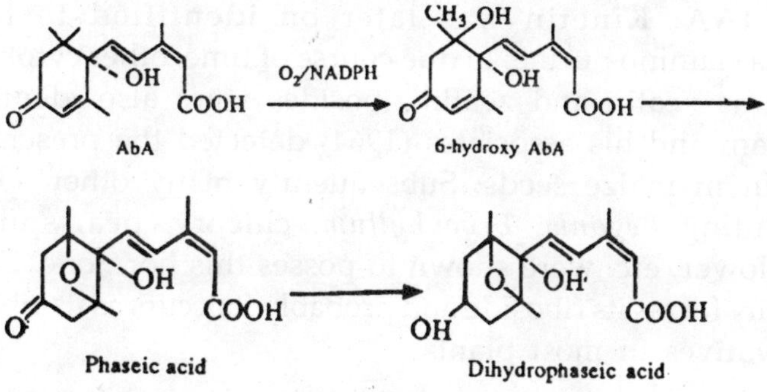

O_2/NADPH

AbA → 6-hydroxy AbA →

Phaseic acid → Dihydrophaseic acid

THE CYTOKININS

Cytokinins is a class of plant hormones which are primarily involved in inducing cell division in parenchymatous cells. The discovery of cytokinins, dates back to 1921, when an Austrian Plant Physiologist, G. Haberlandt discovered that vascular tissues of various plants contained an unknown diffusible factor, which stimulated cell division or cytokinesis. In 1940's, J. Van Overbeek found that milky endosperm of immature coconut fruits also contained this factor, which stimulated cell division and differentiation in very young *Datura* embryos, which did not contain this factor. Studies with many other plant tissues such as *Datura* ovules, yeast, wheat germ and almond meal demonstrated that this factor was of widespread occurrence.

In 1950's, experiments conducted primarily by Folke Skoog, F.C. Steward and their associates identified the chemical factor which was responsible for inducing cell division in tissue cultures. Folke Skoog and his associates while examining the growth of tobacco pith culture, observed that the culture media had to be supplemented with purine base, adenine, to create large callus mass from the cells. The partial breakdown product of yeast or Herring sperm DNA also stimulated cell division and callus formation.

The active compound was named as 'kinetin' which was the first cytokinin to be named. Kinetin and IAA were found to have interactive effects on cell division and differentiation in tobacco pith cultures similar to those shown by adenine and IAA. Kinetin was later on identified to be 6-furfurylamino purine. In due course of time, other cytokinins such as zeatin and zeatin ribosides were also identified. Letham and his associates (1964) detected the presence of zeatin in maize seeds. Subsequently many other species including *Begonia, Bryophyllum,* chicory, pea, spinach, sunflower, etc. were shown to posses this hormone. Zeatin readily forms its riboside and probably it occurs as its riboside derivatives in most plants.

The generic name 'cytokinin' is now used for chemical substances which can stimulate cell division or cytokinesis. However, they are known to affect many other aspects of plant growth and development, as well.

Occurrence and Structure

Several hundred endogenous and synthetic cytokinins have been identified. Most endogenous cytokinins are purine (adenine) or purine derivatives. Zeatin, the most active of naturally occurring cytokinins is 6-(4-hydroxy 3-methyl but-2enyl) amino purine. It has been detected in sweet corn kernels and many other plants. Zeatin exists normally as ribotide (base + ribose + phosphate) derivatives and as such it is present in RNA, specially in the t-RNA of many higher plants and bacteria. About 10 other cytokinins which are adenine derivatives having a side chain on N^6 position have been identified. They may exist either as free base or as ribotides.

Free cytokinins exist in brown and red algae, diatoms, many angiosperms and a few gymnosperms. But they have not been found in mosses and ferns. Certain fungi and bacteria also contain cytokinins that might influence the disease processes caused by these microbes. As a component of t-RNA, cytokinins are present in probably all plants. Cytokinin isopentenyl adenine (I^6Ade) has been identified in t-RNA fractions of yeast, *Escherichia coli, Corynebacterium fascians* and liver cells. It is now well established that the cytokinin is present as the odd base (3' side) adjacent to the anti-codon of all t-RNA molecules that bind to those codons of m-RNA in which the first code letter (or base) is U (uracil). Perhaps this location of the cytokinin in t-RNA assists in amino acid binding and transfer during protein synthesis. Nucleosides of I^6Ade have also been discovered in t-RNA of several systems, but nucleotide form of this cytokinin has not been found.

Bioassay

Plant physiologists have tried several bioassays to detect the presence of cytokinins in plant tissues. However, the most sensitive and specific bioassay is plant callus tissue culture. If the pith from tobacco, soybean and many other dicots is cultured in the presence of auxins, it forms a mass of loosely arranged polyploid cells called *Callus.* However, if cytokinins or an extract suspected to contain cytokinin is also added with auxin; cytokinesis is also promoted and there is an organised growth of the tissue. Besides, stimulation of cytokinesis and differentiation, increased callus growth is also observed with the exogenous addition of cytokinins. In tobacco stem pith callus, the increase in growth is linear in a cytokinin concentration range of 1 and 15 $\mu g\ l^{-1}$. If the Cytokinin to auxin ratio is maintained high, some of the callus cells eventually develop into buds, leaves and stems.

Another bioassay of cytokinins is the lettuce seed germination test. Lettuce seeds have very little or no germination in dark. Red light promotes the germination of seeds. Cytokinin treatment to the seeds also promotes seed germination. However, this bioassay is not very specific, as some gibberellins are also known to promote lettuce seed germination in dark.

A very convenient bioassay of cytokinins is expansion of cotyledons in some dicots, such as radish, cucumber, lettuce, pumpkin, etc. When seeds germinate in dark, the cotyledons are small and yellow. If they germinate in light, growth of cotyledons increases greatly. If the cotyledons from dark germinated seeds are excised and incubated with a cytokinin, growth is enhanced relative to controls without the hormone. Auxins and gibberellins do not have this promotive effect and therefore, the bioassay is specific for cytokinins.

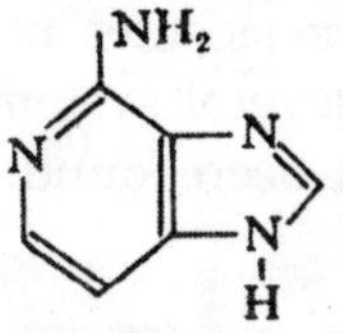

adenine

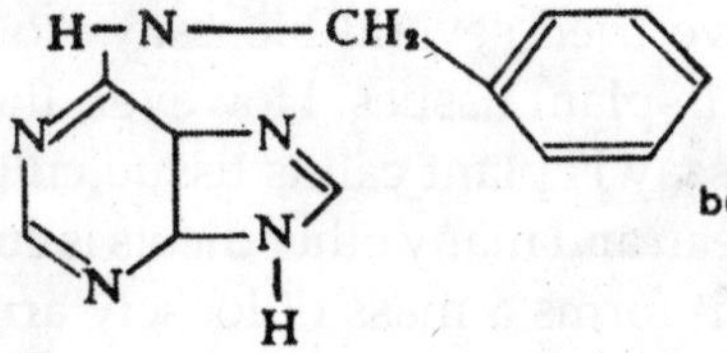

benzyl adenine

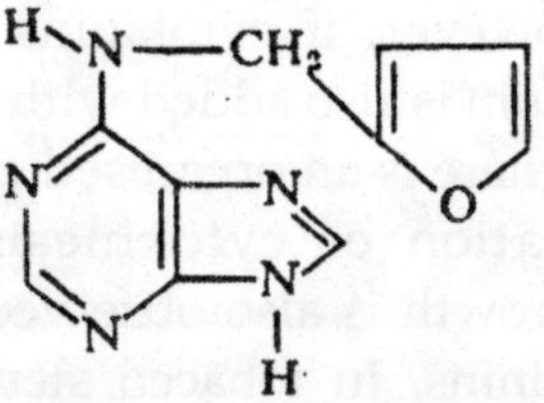

G-furfuryl aminopurine
(kinetin)

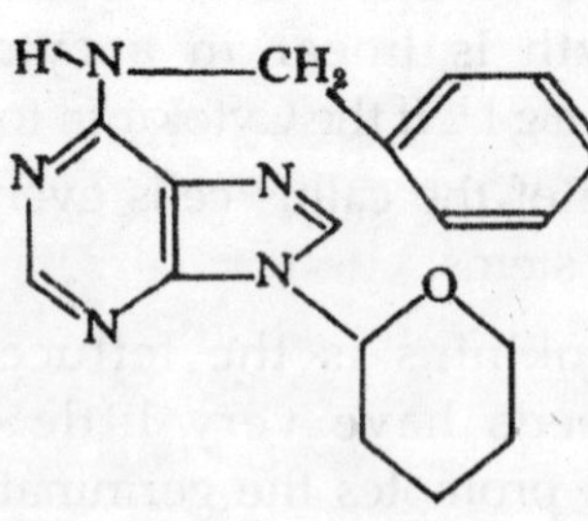

tetrahydropyranyl benzyl adenine

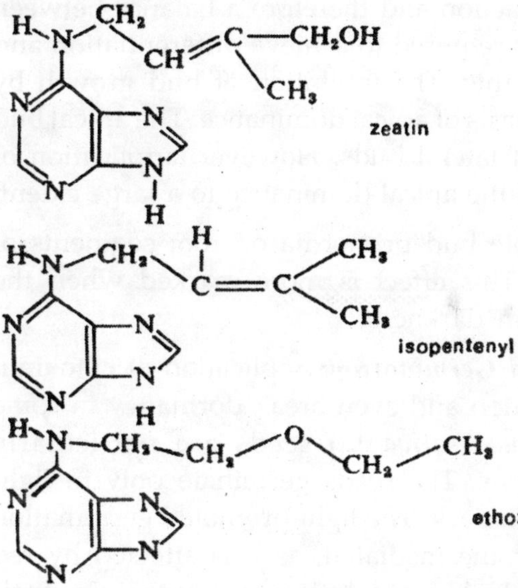

zeatin

isopentenyl adenine or 6 (dimethyl allyl) amino purine

ethoxy ethyl adenine

Structure of some cytokinins

Prevention of *Xanthium* and barley leaf senescence has also been used as bioassay of cytokinins. Leaf discs from light grown *Xanthium* plants are floated on either distilled water or various concentrations of a representative cytokinin in dark. The chlorophyll content is measured after 48 h. The value of chlorophyll as compared to 0 h (immediately after obtaining the leaf discs from intact leaves) is plotted. Cytokinins prevent the decline in chlorophyll content relative to their concentration.

Physiological Effects

Promotion of Cell Division and Organ Formation: A major function of naturally occurring cytokinins is to promote cell division. In the presence of auxins, nearly all cytokinins stimulate cell division and subsequent callus growth in the parenchymatous cells from several plants. By manipulating the ratio of cytokinin to auxin, plantlets can be developed from the tissue in culture.

Cytokinins stimulate bud formation on intact tissues such as leaves, roots, cotyledons or stem pieces as well. Auxins inhibit this cytokinin action and therefore a balance between auxin and cytokinin is required to achieve differentiation and growth of buds in plants. The inhibition of bud growth by auxin is the chemical basis of apical dominance. The apical bud inhibits the growth of lateral buds. However, application of cytokinins relieves off the apical dominance to a large extent.

Cytokinins promote bud primordia on root segments of *Convolvulus arvensis*. This effect is more marked when the segments are grown in darkness.

Promotion of Seed Germination: application of cytokinin can promote germination and even break dormancy in some seeds. Lettuce (*Lactuca sativa* L.) seeds are particularly responsive to cytokinins. The seeds germinate only in light and fail to germinate in dark. The light promoted germination seems to be phytochrome mediated, as it is affected by red light. If seeds are soaked in a solution of a cytokinin, their germination is greatly enhanced in dark, as compared to those seeds which are not soaked in cytokinin. The percentage of germination of cytokinin treated seeds in dark, is similar to those that receive red light treatment.

Expansion of Cotyledons and Leaves: Cytokinins induce expansion of excised cotyledons in several dicots including cucumber, lettuce, mustard, radish, pumpkin, sunflower, etc. Most of these species contain fats as major food reserves in cotyledons. Furthermore, the cotyledons normally emerge above ground and become green and photosynthetic. No response has been found in pea cotyledons which are rich in starch and remain underground after germination and in bean which do not become leafy.

Cytokinins cause cotyledon expansion by increasing cell size, although increased cell division may also account for some expansion. About 10^{-6} M benzyl adenine can increase size of *Xanthium* cotyledons by almost five times.

In lettuce and a few other seeds, where cytokinins stimulate seed germination, the process seems to be mediated via cotyledon expansion. Swelling of cotyledons helps to force the radicle through restrictive seed coat or endosperm barriers and thus the germination is accomplished. Cytokinins stimulate growth of leaf and also cause the expansion of leaf discs in some dicots. 50 μm kinetin causes 2.5 times increase in the diameter of bean leaf discs during a 48 h treatment in dark, as compared to water floated leaf discs in dark. This increase is similar to that caused by red light.

Cytokinins induce another kind of response in excised *Amaranthu* cotyledons. They increase betacyanin synthesis substantially in the excised cotyledons in light. The anthocyanin content increases gradually with incubation time, which is many time higher in cytokinin than in water incubated cotyledons.

Delaying of Senescence: Senescence of leaves is characterised by the breakdown of proteins chlorophylls, nucleic acids and lipids and outflow of the products to other younger tissues. The senescence is often accompanied by yellowing, as chlorophyll is degraded. The senescence or yellowing is accelerated if the leaves are excised from the plant and kept in dark.

A.C. Chibnall first demonstrated in 1954 that the formation of roots in many detached dicot leaves retarded the onset of senescence of the leaves. He proposed that the roots produced some hormones which delayed senescence of such leaves. Subsequently, it was shown by A.E. Richmond and A. Lang (1957) that cytokinin treatment extended the life span of detached leaves by delaying chlorophyll and protein degradation.

Cytokinin mediated delaying of senescence is linked with its effect on translocation of metabolites. The cytokinin treated leaves withdraw nutrients from the adjoining leaves. Thus, while they are able to maintain their nutrition and vigour, the adjoining leaves begin to senesce first. Studies with many dicots and monocots show that if only a part of leaf is treated, radioactive metabolites (such as $^{14}CO_2$) added to other part of the leaf or to the adjacent leaf migrate into and accumulate in the cytokinin treated area of the leaf.

The role of cytokinin induced translocation of metabolites in natural growth and morphogenesis of plants is not clearly understood. However, some investigators have suggested that the process may be involved in the formation of flowers and development of newly folded leaves, which tend to accumulate cytokinin and necessary nutrients.

During flowering, the accumulation of cytokinins in floral buds might cause the preferential transport of the nutrients from the vegetative regions of the plants to the developing floral parts. It is pertinent to mention that, in maize seeds there is a marked increase in cytokinins after fertilization and subsequent development of the kernels. The nutrient mobilization effect of cytokinins is involved in perhaps cytokinin regulation of bud growth and regulation of apical dominance, as well.

Promotion of Chloroplast Development: Exogenously applied cytokinins promote chloroplast developments in callus tissues and excised cotyledons. Tobacco callus tissue, developed in dark, contain etioplasts devoid of chlorophylls with no grana and lamellae. Application of cytokinin in dark promotes lamellar development and if light is also applied simultaneously (with cytokinin) grana and chlorophyll also appear. If no cytokinin is added to light grown callus, development of strorna lamellae, grana and chlorophyll is very low. Thus, cytokinin is essential for the transformation of etioplasts to chloroplasts.

In excised radish and squasn cotyieaons also, eytokinins induce stroma lamellae formation in the etioplasts in dark. When the cotyledons are transferred to light, grana and chlorophyll also appear, as in tobacco callus.

Effect on Overall Plant Growth: Although normal growth and morphogenesis in most plants, required free cytokinins, exogenously applied cytokinins fail to evoke any response in many plants. This is possibly because of a significant pool of endogenous cytokinins. Thus, when the source of endogenous cytokinins (*i.e.,* root) is removed, the remaining plant part responds positively to cytokinin application in most cases. The growth of sunflower and pea shoots can be restored by the application of exogenous cytokinins. Further, in flooded (flooding may retard root growth) tomato plants, foliar application of benzyl adenine, relieves most of the flooding symptoms.

However, in some cases cytokinins inhibit growth of excised shoots and roots. For example, auxin induced elongation of soybean hypocotyls is completely inhibited by kinetin treatment.

Mechanism of Action

The most important aspect of cytokinin action is increased mitosis (cell division) which involves increased synthesis of nucleic acids and proteins. However, some other physiological effects are not easily explained through such a mechanism. Some possible mechanisms of cytokinin action are as follows :

Increased Nucleic Acid and Protein Synthesis: All naturally occurring cytokinins are purine derivatives. This chemical structure relationship between cytokinins and nucleic acids has prompted researchers to suggest that cytokinin action in some way is exerted through influencing nucleic acid and protein metabolism. The nucleic acid metabolism can be influenced in a variety of ways. Some cytokinins are known to intercalate with DNA. Such an interaction will modify the template action of DNA and hence its replication and transcription.

Increase in nucleic acid and protein content during cytokinin treatment has been reported in several systems. In developing rice grains and in leaves of tobacco, kinetin causes marked increase in DNA, RNA and protein. This stimulation of biosynthetic activities in leaf discs, has been suggested to be a factor in the deferral of senescence by kinetin. This cytokinin action is inhibited by the inhibitors of RNA and protein synthesis.

In flax cotyledons and in tobacco pith, during promotion of cytokinesis y cytokinins, there is an increase in nuclear size. This suggests an increase in r-RNA synthesis, a primary function of the nuclei. However, this effect is seen only after several hours and it is likely that the effect is secondary.

In cytokinin induced expansion of cotyledons also increase in RNA has been observed. In peanut cotyledons, all fractions of nucleic acids are stimulated by benzyl adenine treatment.

Cytokinin treatment may induce protein synthesis by increasing one or more aspects of transcriptional and translational activities. These are:

(1) Cytokinins may increase m-RNA and t-RNA content, which are involved in protein synthesis. As described in the earlier paragraph, cytokinins increase nearly all components of nucleic acids in many systems.

(2) Cytokinins are known to bind with some specific ribosomal proteins. This may increase the translational efficiency of ribosomes.

(3) Cytokinin isopentenyl adenosine or its hydroxylated derivative, zeatin, is located as a modified nucleoside, adjacent (3' end) to anticodon in plant t-RNA. In this form, they may affect the covalent linking of charged t-RNA to m-RNA during protein synthesis. In fact, F. Fittler and R.H. Hall (1966) have observed that the presence of a cytokinin adjacent to anticodon in seryl t-RNA from yeast, is essential for protein synthesis in that system.

Increased Enzyme Activity: Several enzymes are known to increase during cytokinin treatment. These enzymes include, those catalysing a variety of chemical reactions of diverse nature. Some examples are, nitrate reductase, tyramine methylpherase proteases, nucleuses, α-amylase, some photosynthetic enzymes and some enzymes of thiamine synthesis. An increased enzyme activity may be the result of increased nucleic acid and protein synthesis.

Increased enzyme activity in some cases however, may be the indirect effect of cytokinins. Increase in nitrate reductase activity by cytokinin has been reported in several systems. In *Agrostemma githago* embryos, the increase with benzyl amino purine is more than that caused by nitrate. This increase seems to be mediated via increased ethylene biosynthesis. Aminoethoxy vinyl glycine, an inhibitor of ethylene biosynthesis, inhibits the cytokinin induced increase in nitrate reductase activity. Further, the ethylene overcomes this inhibition by aminoethoxy vinyl glycine. The increase in nitrate reductase activity by cytokinin, in any tissue will increase overall assimilation of nitrate and long term growth of plant.

***Involvement of a Receptor Protein and a Second Messenger*:** Besides, specific protein binding sites in ribosomes, a protein from tobacco and soybean nuclei has also been isolated, which mediates the kinetin stimulation of RNA synthesis. It is proposed that such a protein serves as the site of cytokinin action and increased DNA, RNA synthesis is due to the action of this hormone-protein complex.

The binding with a receptor protein may also evoke the release or activation of a second messenger molecule. As with other plant hormones, the second messengers may be Ca^{2+}, calcium binding protein calmodulin or polyamines. The second messenger in turn may evoke other biochemical responses.

Cytokinin Metabolism

Biosynthesis: Cytokinins are synthesized in the roots from where they are transported to the shoots via xylem tissues. Biosynthesis of purine type of cytokinins is more clearly known than of other types. Various components of a purine are derived from several small molecules including formyl tetrahydro folate, aspartate, CO_2, glutamine, glycine and methylidyne tetrahydrofolate. The biosynthesis of purine nucleotides (purine base + ribose sugar + phosphate) begins with the ribose phosphate and the purine ring is built on it, step by step. The first nucleotide to be synthesized by this process is inosine monophosphate (IMP). IMP is then converted to either guanosine monophosphate (GMP) or adenosine monophosphate (AMP).

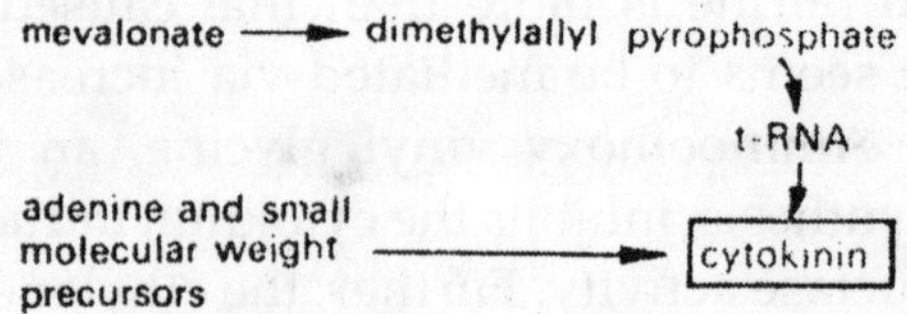

Two alternate routes of cytokinin synthesis

The five carbon side chain attached either to the purine base or to purine nucleotides, in most cytokinins is derived from isopentenyl pyrophosphate. Isopentenyl pyrophosphate is derived from mevalonate utilizing the usual pathway of isoprenoid or steroid biosynthesis. It is believed that the adenosine nucleotide (AMP) is incorporated into t-RNA first and then it is modified; a side chain is added. Using tobacco tissue cultures, C. Chen and R.H. Hall (1969) have demonstrated that mevalonate is incorporated into dimethylallyl and then there is an enzyme system capable of introducing dimethylallyl side chain into the t-RNA . Alternatively, processed nucleotides (as cytokinins) are synthesized first and then incorporated into t-RNA.

Conjugation of Cytokinins: There are some reports to suggest that cytokinins may be inactivated by binding with some other molecules and thus by forming a cytokinin conjugate. For example, glucose has been found to be conjugated to the seventh carbon position (nitrogen atom) of zeatin forming a glucosylzeatin named *raphanatin.* This conjugate is abundant in radish. In some other cases, glucose molecule may also be conjugated to either 9th carbon position or to the hydroxyl group on N-6 substituted side chain.

Degradation of Cytokinins: Destruction of cytokinin is more important than its biosynthesis in regulating the endogenous level of hormone. The important degradation involves the removal of side chain; the nucleoside or nucleotides without the side chain are not active as hormone. This degradation is catalysed by the enzyme cytokinin oxidase activity;

$$N^6 \text{ isopentenyl adenine} \xrightarrow{\text{Cytokinin oxidase}} \text{Adenine + isopentanal or adenosine}$$

The enzyme catalyses the oxidative degradation of cytokinins bearing unsaturated isoprenoid side chain.

Cytokinin oxidase activity is widely distributed in plant tissues. It may exist in free form; although it appears that in bean callus cultures it exists in the form of a glycoprotein.

THE GIBBERELLINS

The discovery of the plant hormone, gibberellins is associated with a disease of rice called "Bakane" (meaning foolish seedling). During the late 1800's, Japanese farmers reported that rice plants affected with this disease were taller and paler and they did not bear seeds. The disease was found to be caused by the fungus *Gibberella fujikuroi* also known as *Fusarium moniliforme.* In 1926, E. Kurosawa found that a chemical from the fungus parasitizing rice seedlings, was responsible for the disease; as sterile filtrates of the fungus caused symptoms of "Bakane" disease in otherwise normal seedlings. Finally, in 1939, T. Yobuta and T. Hayashi isolated a small quantity of highly active crystalline material from the filtrates of the fungus. This material was given the name gibberellin A, as it was isolated from *Gibberella.*

Subsequently they isolated another compound of similar nature and named it gibberellin B. In 1954, British chemists P.A. Brian and others isolated, and chemically characterised a pure compound from culture filtrates of *Gibberella fujikuroi.* They called this new substance gibberellic acid. Almost simultaneously workers in U.S.A. also isolated gibberellic acid from the filtrates of the fungus.

Now several forms of gibberellic acids (gibberellins) have been detected not only in fungi but in uninfected healthy plants of almost all species. A number of gibberellins have been isolated and characterised from higher plants.

Chemical Structure

At present about 60 gibberellins have been identified. They are abbreviated as GA_1, GA_2, GA_3, GA_4 and so on. The original gibberellic acid is GA_3. All the gibberellins are able to promote either stem elongation or cell division or both in plants, although their relative effectiveness may be different. All gibberellins have similar structure. They have a 19 or 20 carbon skeleton derived from terpenoids and grouped in four or five ring systems. This skeleton is called 'gibberellane'. Different gibberellins differ from one another mainly in the types, number and positions of substituent (particularly hydroxyl group) on the ring system and the degree of saturation of 'A' ring. They all seem to have a carboxylic group, and because of this they are referred to as gibberellic acids.

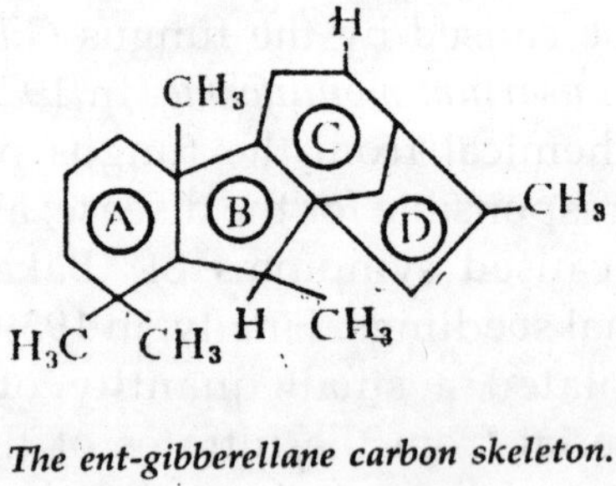

The ent-gibberellane carbon skeleton.

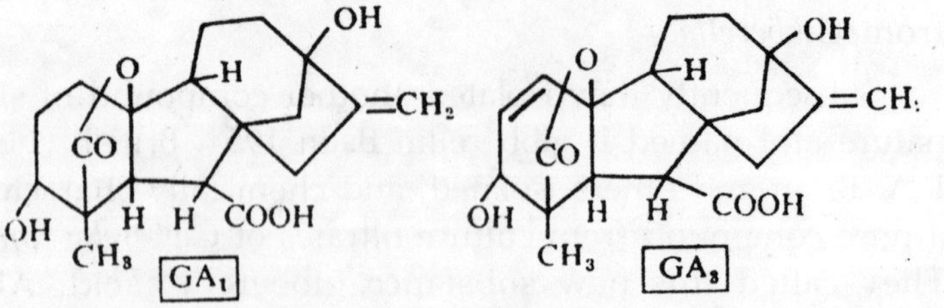

GA_1 GA_3

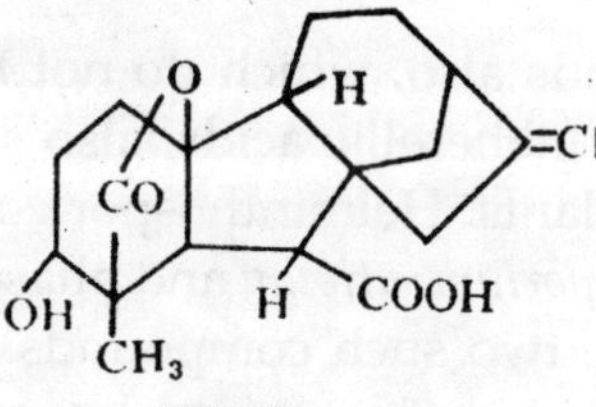

GA_4 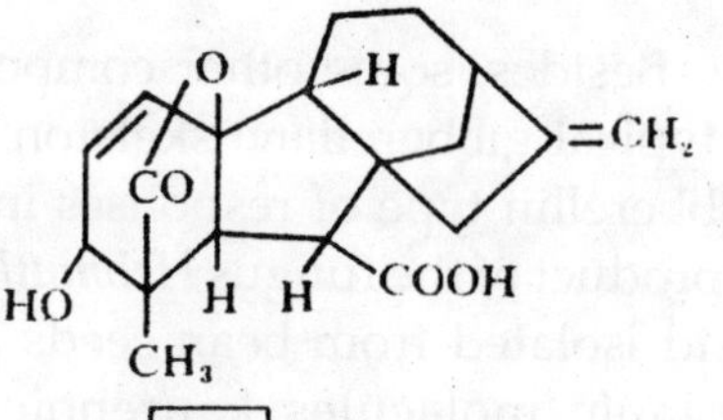 GA_7

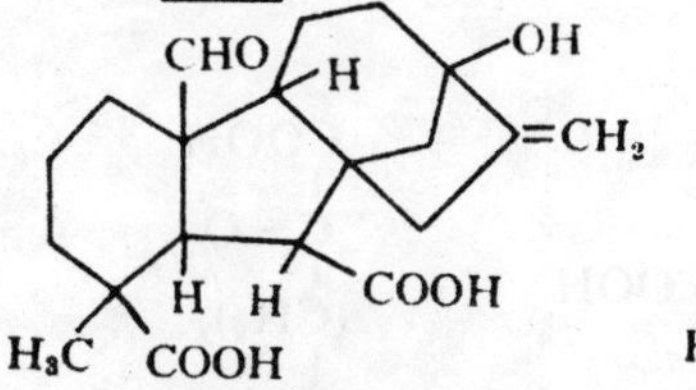

GA_{19} 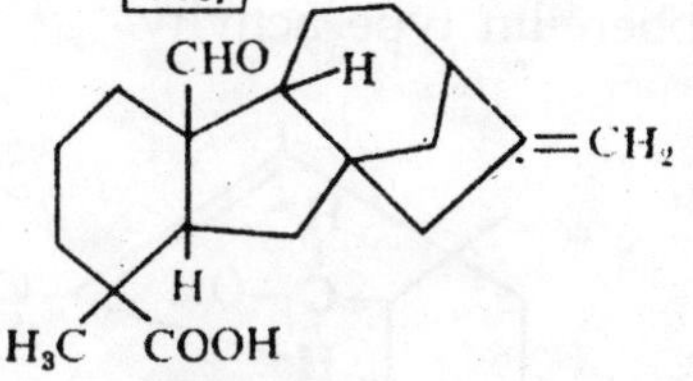 GA_{24}

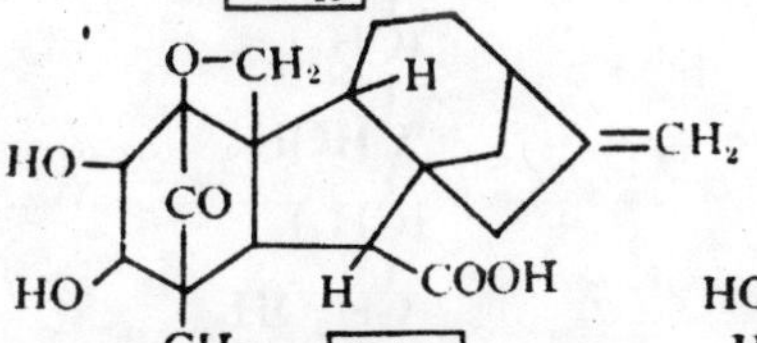

GA_{27} 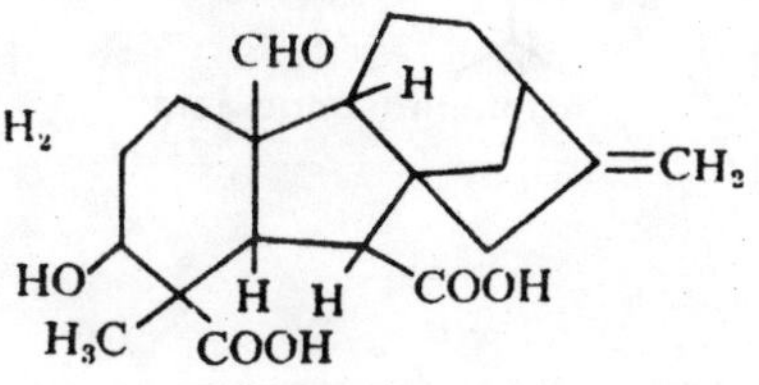 GA_{36}

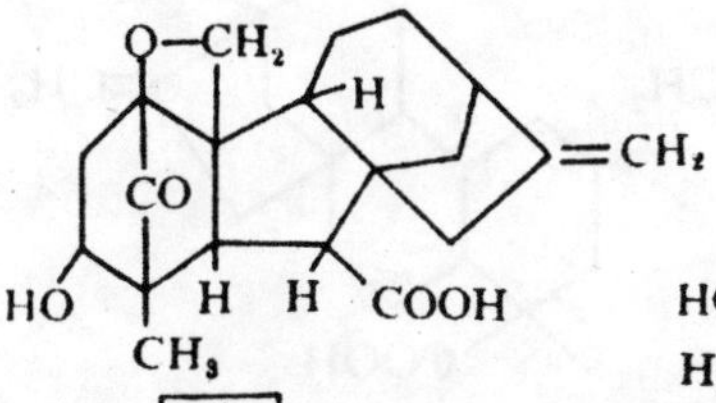

GA_{37} 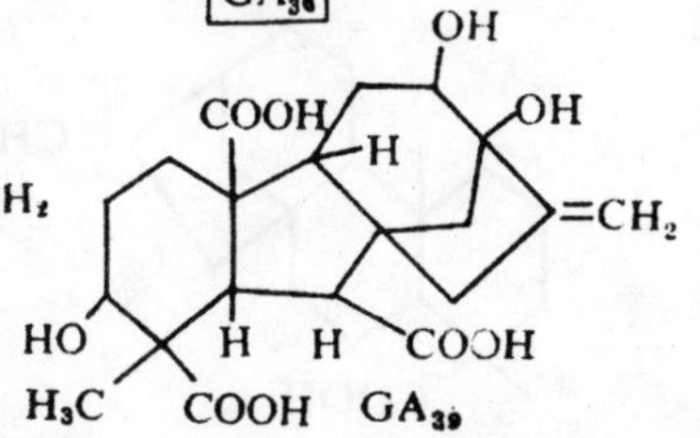 GA_{39}

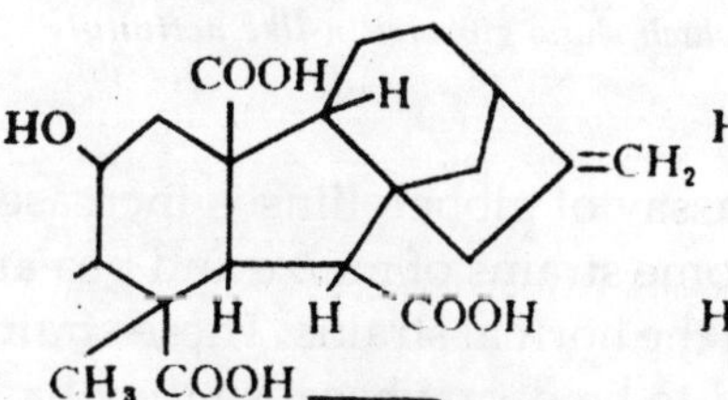

GA_{46} 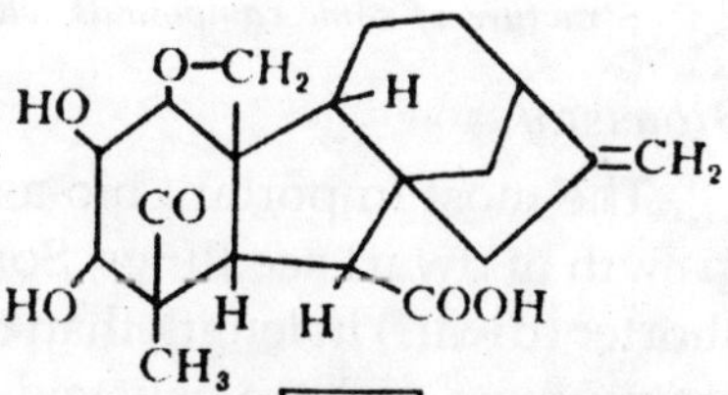 GA_{52}

Structure of some naturally occurring gibberellin.

Besides, some other compounds also, which do not have a typical gibberellane skeleton of gibberellic acids, also show gibberellin type of responses in plants. Helminthosporic acid, a product of the fungus *Helminthosporium sativum* and phaseolic acid isolated from bean seeds are two such compounds. The organic molecules kaurenoic acid and steviol also show gibberellin type activity.

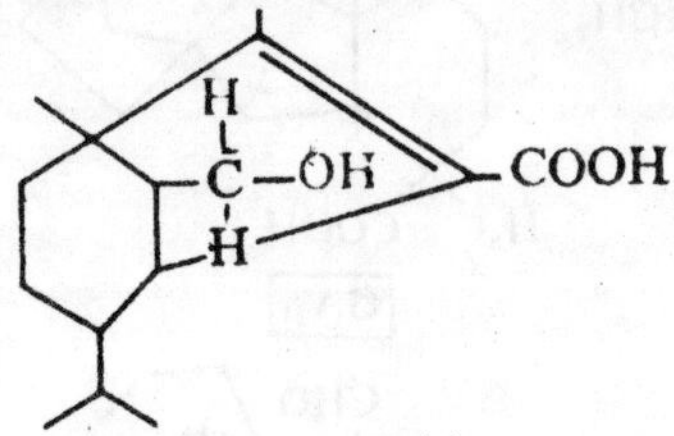

helminthosporic acid

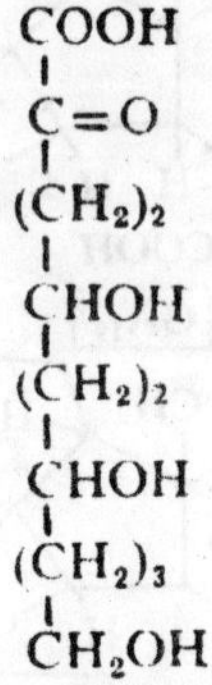

phaseolic acid

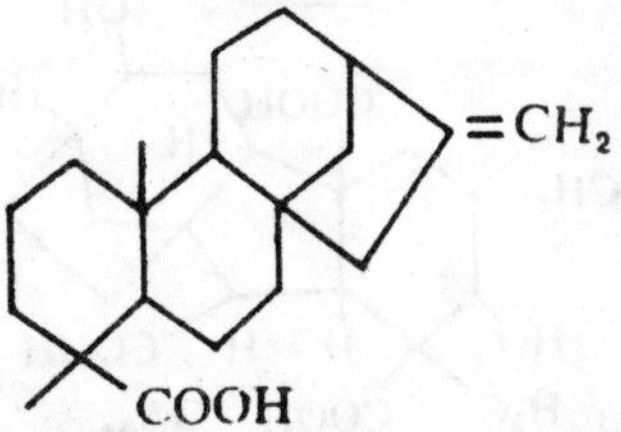

kaurenoic acid

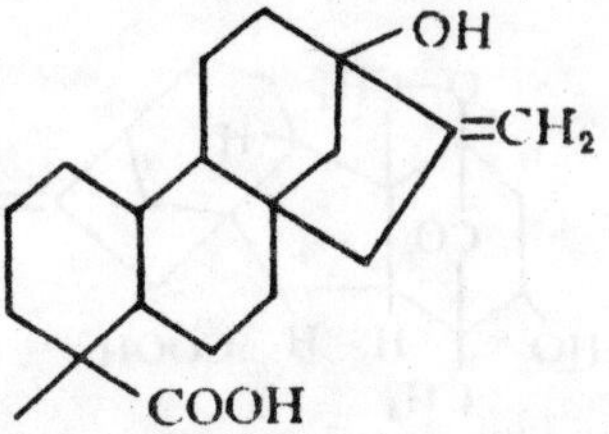

steviol

Structure of some compounds which show gibberellin like activity.

Bioassay

The most important bio-assay of gibberellins is increased growth of dwarf seedlings. Some strains of maize and pea are shorter (dwarf) in length than the normal strains. These strains are mutants and are believed to be dwarf because they have negligible amount of endogenous gibberellins. Application of exogenous gibberellins overcomes the dwarfness of the seedling by increasing the length of the stem. In dwarf maize seedlings, gibberellin solution is applied to the ligule of the first leaf. This causes marked increase in the length of next node or leaf sheath, in a few days. This bioassay is sensitive to as little as 1.0 nanogram of GA. The hormone is applied to the dwarf seedling and increase in epicotyl length is measured after 4-5 days.

Another widely used bioassay of gibberellins is the *barley endosperm test*. By this method, as little as 0.2 nanogram of GA can be detected within 24 h. Barley aleurone (de-embryonated half seeds) layers are incubated in a solution of GA for 24 h. The gibberellins stimulate amylase activity which causes hydrolysis of starch to glucose. The glucose is secreted out in the incubation medium. This accelerated secretion of glucose (reducing sugar) in the incubation medium demonstrates the presence of gibberellin in the test solution.

Bioassays are primarily designed to detect the presence of hormone in plant tissue extract. However, now a days more sophisticated biochemical techniques are available to detect and quantitate the plant hormones. For example, gibberellins can be determined by using gas chromatography, high performance liquid chromatography (HPLC) and other such analytical techniques.

Physiological Effects

In many physiological effects, gibberellins are similar to those of auxins. They also resemble cytokinins in some respects. Some general effects of gibberellins are described in the following paragraphs.

Promotion of Plant Growth: Gibberellins stimulate extensive growth in intact plants. They enhance elongation of intact stems much more than that of excised stem segments. In this respect, their effects are opposite to those of auxins. Because of this property gibberellins are able to overcome genetic dwarfness in some species such as maize, bean and pea. In maize, dwarf varieties are about one fifth in height of the normal varieties. It is believed that dwarfness in such plants is due to a gene mutation, resulting into blocked gibberellin biosynthesis. Lack of gibberellins causes shortening of internodes and reduced height of the plant. Exogenous application of gibberellins fulfils the GA requirement and the plant achieves normal height. Gibberellins have little effect when they are applied to plants of normal height.

The elongation growth of internodes of intact stems depends upon the concentration of GA, and above certain concentration the hormone has no effect.

Gibberellins seem to have a role in dark stimulated increase in plant height, as well. Most plants grow taller in dark than in light, provided other nutritive factors are not limiting. Many workers believe that light inhibition of stem elongation is due to decreased GA level or activity, which in turn decreases the plasticity of cell walls. J.A. Lockardt (1961) showed that the cells of light grown plants were less plastic than those of dark grown. Application of exogenous gibberellins counteracts light induced decrease in plasticity. Plasticity of the cell wall plays an important role in cell elongation land plant-growth.

Promotion of Seed Germination and Bud Growth: Gibberellins promote seed germination in several species, which otherwise fail to germinate unless subjected to low temperature, long days or red light. Gibberellins seem to act as a substitute for these treatments. They enhance cell elongation so the radicle can push through endosperm, seed coat or fruit that restricts its growth.

Subsequently, the hormone also stimulates hydrolysis and transport of stored food materials from endosperm and otyledons to the growing root-shoot axis, specially in the cereals. Gibberellins also overcome dormancy in buds of evergreen and deciduous trees and shrubs. In most cases, emergence from dormancy correlates with an increased level of endogenous gibberellins in seeds and buds.

Induction of Flowering: Exogenous application of gibberellins induces flowering in some photoperiod sensitive and cold requiring species. Gibberellin replaces the requirements both for long day in long day plants and for low temperature in cold requiring species. Applied gibberellins generally do not induce flowering in short day plants except for a few species.

Some examples of species where gibberellins have been found to induce flowering are: *Pharbitis nil, Bryophyllum* spp. *Raphanus sativus, Xanthium* sp., *Impatiens balsamina,* etc. On the other hand, gibberellins inhibit flowering in some woody angiosperms such as apple, *Salix, Bougainvillea,* etc.

Gibberellin changes sex expression of plants, as well. In most species, it induces the formation of male flowers such as in *Cucumis sativa, Cannabis, Bryophyllum,* etc., although in some cases such as in maize it may induce female flowers.

Application of GA increases fruiting and fruit development also in some plants. In tomato, application of GA_3 before flower initiation, increases fruit size. When gibberellins are applied at anthesis to emasculate flowers, parthenocarpic fruits develop. Gibberellins induce parthenocarpic fruits in some other species such as *Prunus,* and cherry as well. But it is not known whether this parthenocarpic effect is mediated via auxin or is independent of that.

Prevention of Senescence: R.A. Fletcher and D.J. Osborne (1965) were first to demonstrate that exogenous application of GA can prevent senescence of leaves. In leaves of *Taraxanum,* kinetin (the principal senescence preventing hormone) has been found to have little effect on senescence, but GA acts powerfully. In the leaves of *Tropaeolum* and *Rumex* both GA and kinetin are active. In the leaves of *Tropaeolum,* it has been also observed that endogenous gibberellin content decreases as the leaves senesce.

The hormone has also been found to delay the senescence of fruits besides that of leaves. Gibberellins are often sprayed on the fruits and leaves of oranges to prevent rind disorders that appear during storage. The hormone delays senescence and maintains firmer rinds.

Mobilization of Food and Minerals in Seeds: During germination of seeds, the organic food and minerals stored in the form of macromolecules in endosperm and cotyledons are released and transported to the growing root and shoot axis, until the seedling is established in the soil.

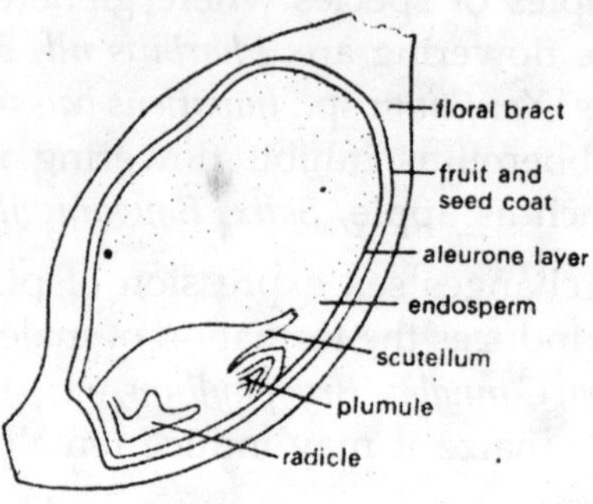

A line diagram of barley seed section to show major tissues.

Thereafter, the seedling starts abstracting nourishment from the soil and atmosphere. Gibberellins are known to stimulate the hydrolysis of stored macromolecules and their transport to the embryonic axis. In cereals, such as in barley, the embryo is surrounded by a thin la; of food reservoir in endosperm. This endosperm layer is called aleurone lay. The aleurone cells provide hydrolytic enzymes that hydroly and digest the starch, proteins, phytin, RNA and certain cell wall material present in the endosperm cells. Gibberellins promote production of the hydrolytic enzymes. In normal conditions, it is assumed that the embryo secretes GA to aleurone layers for increased production of hydrolytic enzymes during seed germination and seedling growth. If the embryo is removed from a barley seed, the aleurone layer does not produce and secrete hydrolytic enzymes. Endogenous or exogenous GA also stimulates the secretion of enzymes out of the endosperm. Reserve mineral elements are also mobilised as a result of GA action.

Mechanism of Action

Gibberellins seem to affect plant growth and development in a variety of ways. In many cases, it has been found that the

GA stimulation of plant growth in dwarf seedlings is associated with increase in auxin content. Further, GA suppresses the activity of Indole acetic acid oxidase, the enzyme responsible for IAA degradation. Based on these observations, it was suggested that GA affects plant processes through auxins. However, several effects of gibberellins are independent of auxins, the auxins can not replace gibberellins. For example, application of GA to barley aleurone layers causes a rapid increase in α -amylase activity, while auxins do not have such effect. It appears that most of the physiological effects of GA are mediated via increased enzyme synthesis and membrane permeability. These aspects are described in the following paragraphs.

Enzyme Induction: Several plant enzymes such as α - and β -amylases, proteases, ribonuclease, RuBP carboxylase, β -1,3-glucanase and nitrate reductase, etc. are known to be induced by gibberellins. J.E. Varner, G.R. Chandra and M.J. Chrispeels in 1960's demonstrated that GA caused *de novo* synthesis of hydrolytic enzymes during seed germination. GA seems to increase both transcription as well as translation, as the increase in enzyme activity is inhibited by actinomycin D (an inhibitor of transcription) and cycloheximide and fluorophenyl alanine (inhibitors of translation). During GA treatment, more of radioactive phenylatanine was incorporated into proteins, indicating an increased translation. There is also an increased incorporation of ribonucleotides into RNA during GA treatment.

In barley aleurone layers, GA specially enhances the translatable m-RNA for α -amylase. The increase in RNA seems to be the result of increased RNA polymerase activity as has been demonstrated in pea stems and wounded potato tubers. The hormone may also unveil some sectors of the chromatin DNA, making it available for RNA synthesis. However, gibberellins are known to inhibit the activity of some enzyme such as invertase and peroxidase in sugar cane.

Membrane Permeability: Some effects of gibberellins seem to be mediated via increased permeability of the cellular and organellar membranes. Gibberellins for example, not only increase the synthesis of α-amylase but also activate its release from barley aleurone layers. In the presence of exogenous GA, a major portion of enzyme activity is detected in incubation medium which increases with the incubation time. H.W. Evins (1971) detected changes in the aleurone cell membranes within 1 to 2 h of GA treatment. The permeability of some synthetic membranes has also been shown to be affected by GA treatment. One synthetic membranal identity is the liposomes, which are prepared from lecithin, cholesterol and diacetyl phosphate. Permeability of liposomes to glucose increased during GA treatment. In this respect, GA is similar to many steroid hormones of animal origin, which also change the membrane permeability.

Gibberellins may affect the permeability of plant membranes either by affecting the synthesis of molecules involved in membranes or by changing their organization. The enzymes choline transferase and phosphatyl choline glyceride transferase increase by 2 to 4 folds after 2 to 12 h GA treatment (Johnson and Kende, 1971). These enzymes are involved in the synthesis of lecithin, a major phospholipid membrane component. Further, GA_3 treatment increases the incorporation of ^{14}C-Choline in partially purified endoplasmic reticulum preparations.

Gibberellin Interaction with DNA: Gibberellins and possibly auxins and cytokinins as well, have been reported to interact with DNA and change its physical properties. GA_3, IAA and kinetic change the melting points of DNA. GA_3 has been shown to bind with adenine-thymine rich segments of DNA in *in vitro* conditions (Kessler and Snir,1969). However, it is not known whether such interactions of GA with DNA are necessary for various physiological and biochemical responses.

Analogy with Animal Hormones: Structural similarities of GA with steroid hormones (of animal origin) have Prompted suggestions that GA may have a mode of action that resembles that of steroid hormones of animals. Accordingly, the hormone may bind to a specific site (receptor) in the membrane of the target cell (the cell where the hormone is to elicit response). There is a structural change in receptor Protein as a result of this binding, which then moves on to the nucleus, binds to the chromatin DNA and activates transcription.

Using another analogy with animal hormone systems, some scientists have demonstrated that cyclic AMP (a messenger molecule in animal hormone action) can stimulate α-amylase formation in barley aleurone layers in a manner similar to GA. The effects of cyclic AMP are inhibited by abscisic acid, cycloheximide and 6-methyl purine as the GA effects are.

However, the presence of a specific hormone receptor and involvement of cyclic AMP as a second messenger molecule do not appear to be universal phenomena.

Anti-Gibberellins

A number of chemical compounds have been identified and synthesized by the chemists during the last three decades, which have effects opposite to those of gibberellins. Since most of these compounds cause dwarfing in plants, they are commonly referred to as growth retardants. Regarding the mode of action of these compounds, it has been shown that they inhibit the biosynthesis of endogenous gibberellins and therefore, when exogenous gibberellins are applied, the dwarfing effect of these growth retardants is overcome. C.A. West and his associates in 1960's demonstrated that growth retardants, AMO 1618, CCC and Phosphon D block the conversion of geranyl geranyl pyrophosphate to copalyl pyrophosphate and therefore inhibit the synthesis of kaurene and other such compounds.

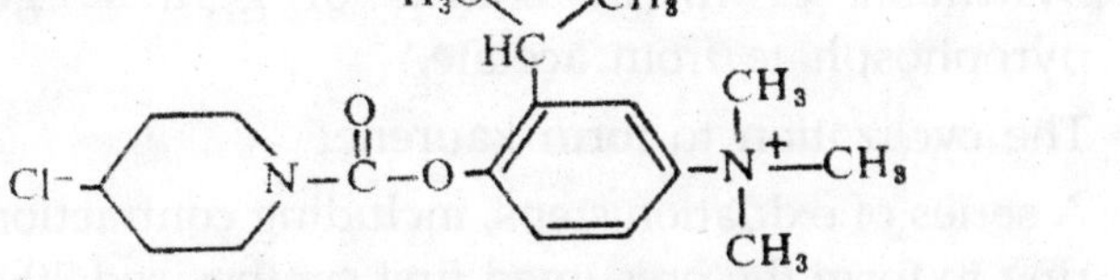

AMO 1618 [2'—iso propyl-4' (trimethyl ammonium chloride) 5' methyl phenyl piperidine carboxylate]

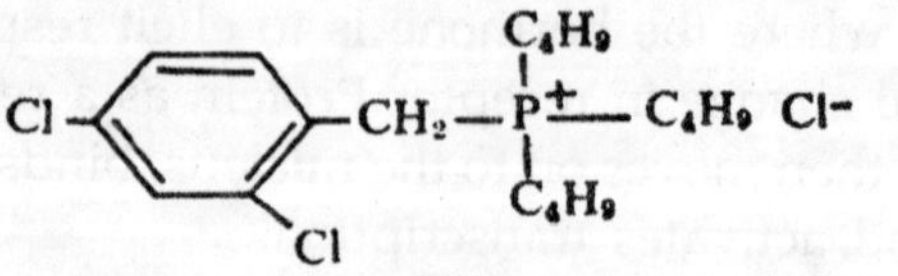

$$Cl-C_6H_3(Cl)-CH_2-\overset{C_4H_9}{\underset{C_4H_9}{P^{+}}}-C_4H_9\ Cl^-$$

phosphon-D [tributyl 2, 4- dichlorobenzyl phosphonium chloride]

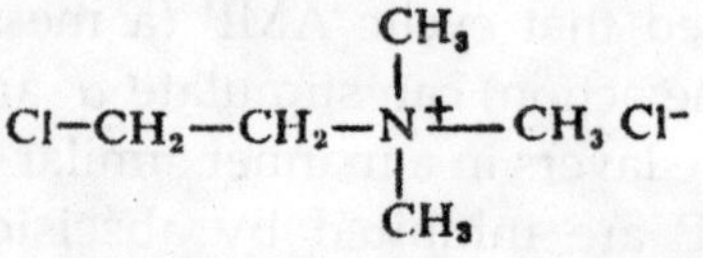

$$Cl-CH_2-CH_2-\overset{CH_3}{\underset{CH_3}{N^{+}}}-CH_3\ Cl^-$$

cycocel, CCC [B-chloroethyl trimethyl ammonium chloride]

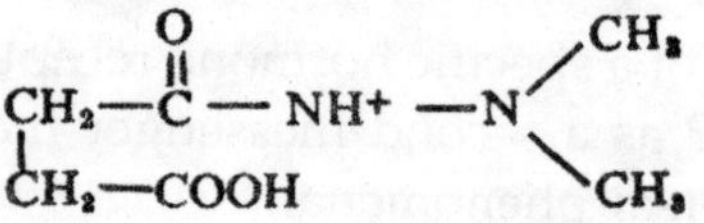

$$\begin{array}{l} CH_2-\overset{O}{\overset{\|}{C}}-NH^+-N(CH_3)_2 \\ | \\ CH_2-COOH \end{array}$$

SADH [succinic acid 2, 2 dimethyl hydrazide]

Structure of some growth retardants

In addition, abscisic acid also inhibits GA activity in several cases. Two diterpenoids epiallogibberellic acid and atractyligenin have also been identified to inhibit GA activity.

Gibberellin Metabolism

Biosynthesis: Biosynthesis of gibberellins has been examined in immature seeds. However, there are evidences in favour of gibberellin synthesis in roots as well. Most of the biosynthetic steps take place inside the plastids. By using tracer technique (use of radioactive isotope) it has been shown that acetate is the primary precursor of gibberellin synthesis. The synthesis of various gibberellins from acetate may be considered to involve four groups of reactions:

(1) Synthesis of linear chains of geranyl geranyl pyrophosphate from acetate.

(2) The cyclization to form kaurene.

(3) A series of oxidation steps, including contraction of 'B' ring to form the presumed first synthesized 20-carbon gibberellin, GA_{12} and the 19-carbon GA_4.

(4) Interconversions of various GAs.

To begin with, acetate molecule is activated by coenzyme A (CoA) and acetyl CoA is formed. In the first three steps, 3 acetyl CoA molecules condense in series to produce mevalonic acid. In the presence of two ATP molecules, mevalonic acid is phosphorylated to form mevalonic acid pyrophosphate. Decarboxylation of this compound in the presence of ATP yields isopentenyl pyrophosphate (IPP). IPP is a 5-carbon compound which is precursor of gibberellins, carotenoids, abscisic acid and a portion of cytokinins. Isomerization of IPP to form dimethyl allyl pyrophosphate is the initial step in the synthesis of higher terpenoids. One molecule of dimethyl allyl pyrophosphate combines with one molecule of isopentenyl pyrophosphate to yield a 10-carbon compound geraniol pyrophosphate. In subsequent two reactions, two molecules of IPP are added to yield a 20-carbon compound geranyl geraniol pyrophosphate.

In the second group of reactions, geranyl geraniol pyrophosphate is first converted to the diterpene alcohol copalyl pyrophosphate and then to kaurene. As indicated earlier, this conversion is inhibited by anti-gibberellins AMO 1618, CCC and phosphon D.

In the third group of reactions, kaurene is converted to 20-carbon GA_{12} and 19-carbon GA_4. These steps involve oxidation and most of the experimental work in this sector has been done with fungal cultures, where the enzyme systems for such conversions are located on microsomes.

$$\underset{\text{acetate}}{CH_3COOH} \underset{\text{thiokinase}}{\overset{\text{COA}}{\rightleftharpoons}} \underset{\text{acetyl COA}}{CH_3CO.COA} \underset{\text{ketothiolase}}{\overset{\text{acetyl COA} \quad \text{COA}}{\rightleftharpoons}} \underset{\text{acetoacetyl COA}}{CH_3CO-CH_2\,CO.COA}$$

$$\underset{\text{synthetase}}{\overset{\text{acetyl COA} \quad \text{COA}}{\rightleftharpoons}} \underset{\text{hydroxyl methyl glutaryl COA}}{CH_3\,C(OH).CH_2CO\,COA \atop |\;CH_2COOH} \underset{\text{reductase}}{\overset{2NADPH \quad 2NADP^+}{\longrightarrow}} \underset{\text{mevalonic adid}}{CH_3-C(OH)(CH_2COOH)-CH_2-CH_2OH}$$

Biosynthesis of gibberellins from acetate.

Some known interconversions of gibberellins.

Interconversion of gibberellins forms the fourth group of reactions in gibberellin biosynthesis. However, these interconversions do not seem to be easily carried out in all Plants.

Gibberellin Inactivation

By using labelled gibberellins it has been demonstrated that the hormone is rapidly metabolised in plants. The inactivation of gibberellins can take place either by its hydroxylation or by forming conjugates with glucose.

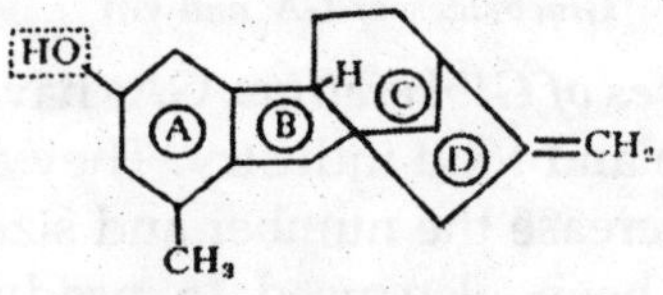

2, β *-hydroxylated gibberellin.*

Addition of hydroxyl group in the B configuration to carbon number 2 on the 'A' ring is a common inactivation process in higher plants. This hydroxylation does not appear to be reversible and results in complete loss of or at least greatly reduced biological activity of the gibberellin. Some examples of known 2 B hydroxylation of gibberellins in plants are:

$$GA_1 \rightarrow GA_8, GA_{30} \rightarrow GA_{29}, \text{ and } GA_9 \rightarrow GA_5.$$

In each case the product has a very low GA activity.

Several types of GA glucosides have been identified in *Pharbitis* and other plants. The highest concentration of these conjugates have been found in mature seeds, and it has been suggested that they may be storage forms of gibberellins which could be hydrolysed to release free gibberellins at the time of seed germination

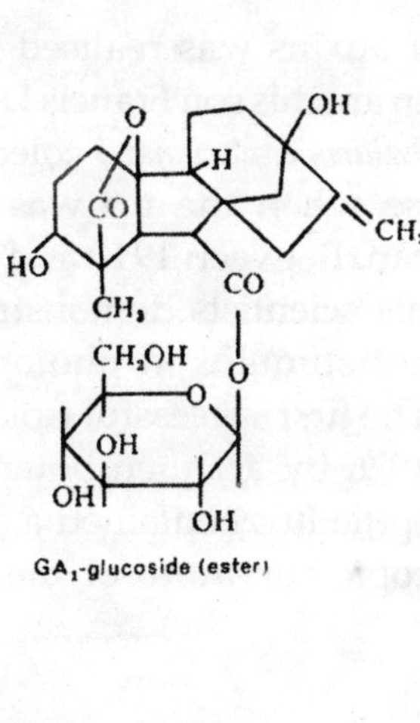

GA_1-glucoside (ester)

GA_8-glucoside (ether)

Glucosides of GA_1 and GA_8.

Commercial uses of Gibberellins: GAs have found extensive use in horticulture and food industry. They are often sprayed on vineyards to increase the number and size of grapes in the cluster. GA_3 has been also used to produce 'Thompson's seedless' grapes. Improvement in the size, colour and quality of apples, pears and many other fruits is also achieved by spraying gibberellins. In these cases, the hormone stimulates bud formation and fruit set.

Gibberellins are also used to increase α-amylase activity in germinating barley seeds which is used for malt production in beer industry.

One of the most notable uses of gibberellins is their application to sugar cane crop. They promote the elongation of cane internodes without decreasing the sugar content. This application is particularly important to Indian sub continent, where sugar cane is an important cash crop and foreign exchange earner.

The Auxins

The existence of auxins was realised as early as in 1887, when Charles Darwin and his son Francis Darwin demonstrated that canary grass *(Phalaris canariensis)* coleoptiles did not show phototropic response when the tip was either removed or covered with a foil cap. Between 1910 and 1930, investigations conducted by various scientists demonstrated that the agent which perceived light stimulus in phototropic response was of chemical nature. The first successful isolation of the chemical was carried on in 1926 by a Dutch botanist, F.W. Went. He found that oat coleoptile tips contained a chemical which was involved in phototropic curvature of the coleoptile.

The chemical was also involved in endogenous control of growth rates. The chemical which was called auxin (from the Greek auxein meaning to grow), by Kogl, Hagensmit and Went, was later on isolated from a variety of materials including human urine (Kogl and Kostermann, 1934), malt and corn meal (Hagensmit *et al.* 1942). Now we know that auxins are present in the tissues of all higher plants, although only in small quantities. They are of fundamental importance in the physiology of growth and differentiation.

Chemical Structure

The first crystalline auxin was obtained from human urine by Kogl and others in 1934. Chemically it was shown to be indole 3-acetic acid (abbreviated IAA);

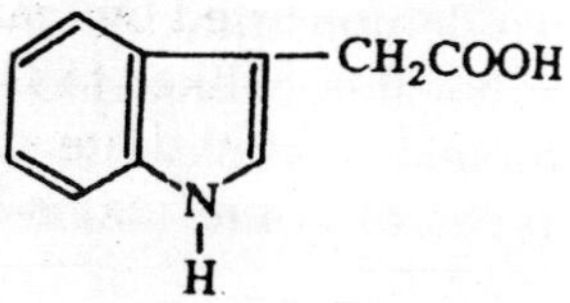

CH_2COOH

N

H

Indole 3-acetic acid

Since than it has been found that IAA occurs as a principal auxin in most plant species. There are many other compounds closely related chemically to IAA, which also possess auxin like properties. Other indole compounds such as methyl 3-acetate, ethyl indole 3-acetate, 4-chloro 3-indole acetic acid, indole 3-acetamide, 4-chloroindole 3-acetic acid, indole 3-ethanol and indole 3-carboxylic acid also possess auxin properties. It is believed that most of these compounds which all have an indole group, are ultimately converted to IAA. There are also compounds such as phenyl acetic acid present in plants, which are not indole and yet possess auxin activity. However, the chemical nature and physiological significance of their auxin like properties are largely unknown. Since more than one type of chemical compound shows auxin properties, we speak of auxins (plural) rather than auxin (singular).

Structure of related auxins other than indole 3-acetic acid.

Name of the compound	Chemical structure	Discovered by
Indole acetaldehyde	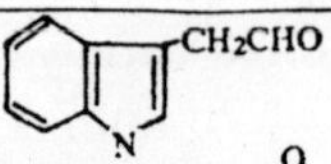 CH_2CHO	Larsen (1944)
Indole pyruvic acid	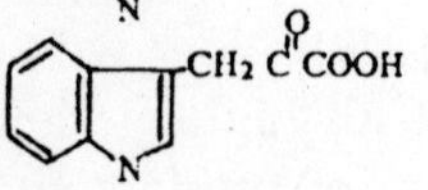 $CH_2\overset{O}{\overset{\|\|}{C}}COOH$	Yamaki and Nakamura (1952)
Indole aceto nitrile	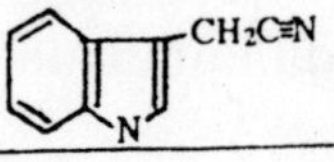 $CH_2C{\equiv}N$	Jones *et al.* (1952)

Bound Auxins

Since the first suggestion of Kenneth Thimann and Folke Skoog (1940), that the endogenous IAA may be bound to some proteins, the existence of bound auxins in plant tissues specially in the seeds, has been demonstrated by many others as well. In bound auxins, IAA is covalently linked to a substituent molecule which may be a protein, carbohydrate or alcohol. In maize kernels, following types of bound auxins have been detected;

(1) Esters of IAA.
(2) IAA conjugated with cellulosic glucan.
(3) IAA-inositol conjugate.
(4) IAA-myoinositol-glycosides conjugate.

Bound auxins are the storage forms of hormone. They are synthesized during seed maturation and are hydrolysed during seed germination. These compounds play important roles in establishing the level of free hormones in plant tissues.

Synthetic Auxins: Several synthetic chemicals of diverse chemical structure also behave as auxins. Some examples are : phenoxy acetic acid, naphthalene acetic acid, picolinic acid, etc. Most of these chemicals have agricultural applications, generally for purposes other than regulating growth of the plants. For example, 2,4-dichlorophenoxy acetic acid is used as a herbicide (Table). Generally, these molecules have following structural peculiarities;

(1) An unsaturated ring.
(2) An acidic side chain.

(3) Some spatial relationship between the unsaturated ring and acidic side chain.

These structural requirements seem to relate to requirements for an attachment to some site in the cell.

Structure of some synthetic auxins.

Name of the compound	*Chemical structure*
2,4-dichlorophenoxy acetic acid	OCH_2COOH, Cl, Cl
Naphthalene acetic acid	CH_2COOH
β-naphthoxy acetic acid	OCH_2COOH
2,4,6-trichlorobenzoic acid	Cl, COOH, Cl, Cl
2,3,6-trichlorobenzoic acid	Cl, COOH, Cl, Cl
4-amino-3,5,6-trichloropicolinic acid	Cl, N, COOH, Cl, Cl, NH_2

Bioassay of Auxins

Detection of a plant growth substance using a biological test material is termed bio-assay. The most widely used bio-assay of auxins is the *Avena coleoptile curvature test,* which has been used since early works on auxins. Unilateral application of solution suspected to contain auxin causes curvature in the coleoptile.

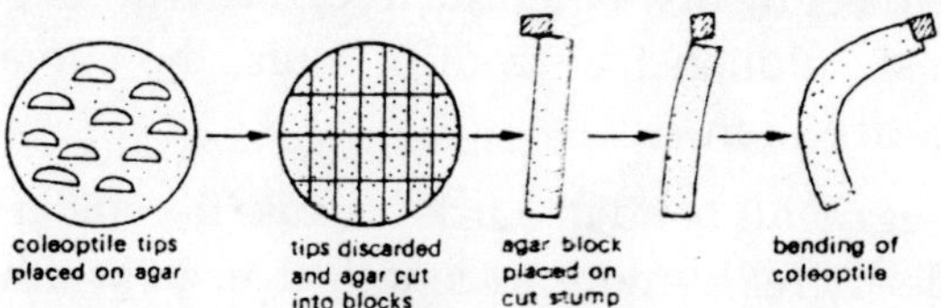

Avena Coleoptile bio-assay of auxins.

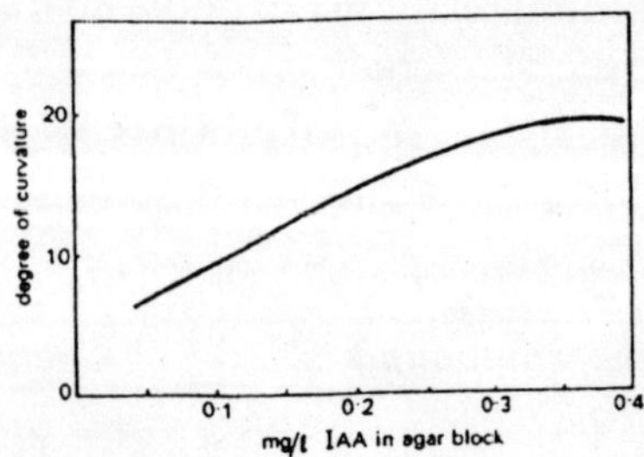

Effect of IAA concentration on curvature of Avena Coleoptile.

The excised tissue (may be a coleoptile tip) suspected to contain auxin is placed on dissolved agar for a few hours, so that the auxin diffuses out to the agar. The agar is then cut into small blocks. Each agar block is then kept on one side of a decapitated coleoptile stump. The set is placed in total darkness. Curvature in the coleoptile shows the presence of auxin in the agar block. The amount of auxin is directly proportional to the degree of curvature. An agar block which had not been in contact with the suspected tissue or auxins is used as a control. When such a block is kept on a decapitated stump, there is no curvature.

Another convenient bioassay of the auxins is the *split pea stem curvature test*. This bio-assay also depends upon the differential growth response of the auxin. A stem section of pea seedling is slit longitudinally and floated on the solution suspected to contain auxin. The solution may be an extract of the plant tissue, which is to be tested for the presence of auxin. The sections are presoaked in distilled water for about one hour to remove any endogenous auxin that might be present in the stem segments. When such segments are slit and floated on the test solution for about 6 hours, the curvature of the segments are seen.

The segments bend inwards because the epidermal cells on outer-side increase more in length than in width, while the cortical cells on the inner-side respond to auxin with more growth in width than in length.

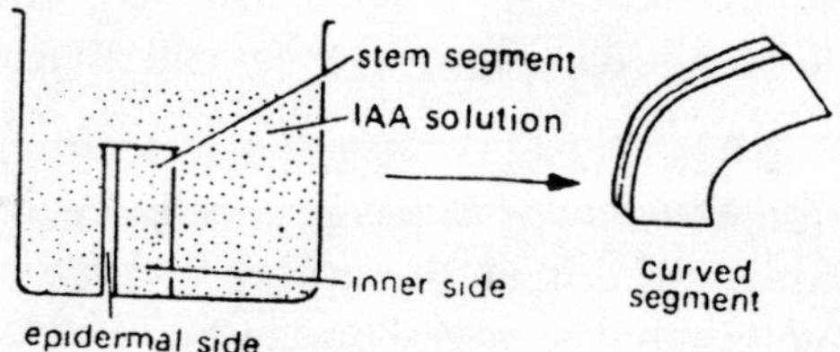

Unequal Growth of two sides of a cut stem segment in IAA solution.

Longitudinal increase in growth of coleoptile or stem segments is also used sometimes as bio-assay of the auxins. When small pieces of coleoptile or stem are immersed in an auxin solution quantitative stimulations of elongation are obtained. The stimulations are roughly proportional to the logarithm of the concentration of the auxin. Etiolated stem segments are more sensitive for such a bioassay than green ones.

Physiological Effects

Auxins affect a variety of response in plants. Some of these are described in the following paragraphs:

1. *Stem Elongation:* Application of exogenous auxins causes stimulation of coleoptile and stem growth. The effect varies according to the tissues. In general, the most marked stimulation are obtained in stems and coleoptiles of etiolated seedlings. The stimulation of elongation is also seen in cut stem segments. The stem segments of etiolated pea and soybean seedlings are often used as bio-assay materials. However, at high concentrations, auxins, inhibit growth instead of causing stimulation. This inhibition is believed to be associated with ethylene accumulation at high auxin concentrations.

2. *Coleoptile Curvature Effect:* As described in the bio-assay section auxins cause bending of the coleoptile. Bending occurs because of the polar transport of auxins. This polarity is much more evident in coleoptile than in many stem tissues. The presence of auxins on only one

side of the coleoptile increases cell division on that side and therefore the coleoptile bends on other side.

3. *Apical Dominance:* It has generally been observed that in most seedlings, the apical tip of the stem inhibits development of lateral buds within certain areas of the apex. If the apical tip is removed, the lateral buds develop. The inhibition of lateral bud formation by the presence of apex is called apical dominance. This inhibition is attributed to the presence of auxins in the apex. If auxin is applied to the cut stump of a seedling, then also growth of lateral buds is inhibited.

4. *Rooting:* The auxin, indole acetic acid has been identified as a rooting hormone. Exogenous application of IAA induces rooting in several stem cuttings. It hastens rooting as well as increases the number of roots formed. In many trees, formation of adventitious roots can be induced by the treatment with auxins. As such, auxins have found commercial application as rooting hormone.

   Elongation of roots also increases with the application of exogenous auxins, although in most cases, it inhibits root growth in intact plants.

5. *Cambial Activity:* Stimulation of cambial activity in trees by exogenous application of auxins was first reported in 1936 by Soding. Now it has been recorded in several other studies. Xylem differentiation is more pronounced than the phloem. In fact only when IAA is present in the medium for tissue culture, no phloem differentiation occurs. The auxins also promote cell elongation and division in the tissues of shoot tips.

6. *Flowering:* The exogenous application of auxins increases formation of female flowers and ovary wall growth leading to parthenocarpy in cucurbits, grapes, etc.

   This effect however, seems to be indirect. Burg and Burg (1966) have reported that auxins induce ethylene production which may be the cause of parthenocarpy.

7. *Abscission:* CD. Larue (1936) demonstrated for the first time that many synthetic auxins inhibited the abscission of coleus leaves. Since then, control of abscission by IAA has been demonstrated in various other plants as well. In some cases, IAA induces abscission, specially in the aged tissues.

   This is believed to be due to auxin induced synthesis of ethylene, as ethylene is known to promote abscission.

8. *Plant Growth Movements:* Because of their polar transport and effect on cellular growth, auxins are known to be involved in phototropic and geotropic plant movements. When a growing plant is illuminated by a unilateral light, it responds by bending towards light. According to Cholodony-Went theory (1927, 1928), there is a higher concentration of auxin on the shaded side than on the light side of a unilaterally illuminated shoot. This unequal distribution of auxins could be the result of light induced inactivation of auxins on light side, light induced lateral transport of auxin or inhibition of basipetal transport of auxin. The result of unequal distribution of auxins, is unequal growth of shoot, more on dark side than on light side, resulting in a curvature.

The geotropic movement of root and shoot is also mediated via endogenous auxin levels. The differential growth of a horizontally placed root or shoot is due to the accumulation of auxins on lower side. The accumulation in the stem causes accelerated growth on the lower side and the stem bends upwards (negative geotropism). When auxins accumulate on the lower side of the roots, however, it causes inhibition of cell elongation on that side. The root bends downwards, as the growth on upperside is uninhibited, causing positive geotropism.

Mechanism of Action

Auxins affect a variety of physiological processes in plants, as is evident from the foregoing account. Some of these effects are very fast, while others have a lag of few minutes to few hours. It appears however, that at molecular level and perhaps other plant hormones as well, affect nucleic acid directed protein synthesis, enzyme activity and membrane permeability, either directly or through some other metabolites. All physiological effects are manifestations of one or more of these effects. These molecular aspects are described in the following paragraphs :

Activation of Gene Expression: The relationship between the effects of auxins on nucleic acids and on growth was first suggested by F. Skoog in 1954. Since then several others have also shown that the effects of auxins on cell growth was through nucleic acid and protein synthesis. Auxins seem to be promoting both transcription (synthesis of RNA on DNA) and translation (m-RNA directed protein synthesis) processes. In tobacco callus tissue, IAA induced growth is preceded by a proportional increase in RNA.

Increased incorporation of labelled nucleotides into RNA during auxin treatment, has also been observed. Both, increase in growth and in RNA is inhibited by the antibiotic actinomycin D. The anti-auxins which inhibit auxin induced growth also inhibit auxin induced transcription. Experiments with isolated nuclei or chromatin indicate that auxin's effects on transcription may be mediated via the release of transcription factor from the plasma membrane. This factor is able to stimulate *in vitro* RNA polymerase activity.

Treatment with auxins also stimulates translation of m-RNA in several cases. Treatment of pea epicotyls with auxins increases the membrane bound polysomes. These polysomes are able to synthesize the enzyme cellulase under *in vitro* conditions with the help of cellulase specific m-RNA. Auxin induced cellulase activity is suppressed by protein synthesis inhibitors. It has been suggested by D.P.S. Verma and his associates (1975) that auxins regulate cellulase activity both at transcription and translation level. Increase in transcription or translation will increase the level of either a specific protein (cellulase for example) or total protein pool.

Models for Auxin induced Gene Expression: Protein synthesis in every organism is under the control of DNA directed m-RNA synthesis (transcription) and translation of the m-RNA. As discussed earlier, increase in enzyme (protein) activity by auxins in many cases, is mediated via increased transcription of DNA and translation of m-RNA. Auxins may increase specific enzyme synthesis by increasing the transcription of specific genes (DNA segments). Athanasios Theologis (1986) of St. Louis, U.S.A. has proposed two possible models for auxins mediated regulation of specific genes: (i) Negative control model and (ii) Positive control model.

According to negative control model a short lived proteinaceous repressor molecule (R) attaches to the promoter site, adjacent to the auxin responsive gene(s), namely structural genes. This attachment inhibits the transcription of structural genes. In the presence of auxin, this repressor is inactivated; it does not bind to the promoter and structural genes express themselves.

According to the positive control, an endogenous activator (A) molecule is proposed to function in the regulation of gene expression by auxins. This activator may be under direct control of auxin. It is inactive in the absence of auxin and active in the presence of it. When active, it attaches to the promoter site and causes expression of the structural genes by inducing transcription. Alternatively, the activator may be controlled by the auxin in an indirect manner. The activator in the absence of auxin interacts with a proteinaceous repressor molecule (R), resulting in the inactivity of the former. When inactive, it does not bind with the promoter and hence causes no transcription. However, in the presence of auxins, the repressor is inactivated and then activator is free to induce transcription. It has also been proposed that the activator binds with RNA polymerase and induces transcription.

Increased Enzyme Activity: Auxins may directly modulate the activity of some enzymes without causing any increased synthesis. In pea stem segments, auxin increases the activity of glucan synthetase, an enzyme involved in the cell wall metabolism, possibly through the phosphorylation of inactive form of the enzyme. Activation of RNA polymerase activity, as observed in soybean hypocotyls and lentil roots, also seems to be mediated via physico-chemical modulation of the enzyme. A.J. Gordon and A.E. Flood (1980), by using labelling technique demonstrated that 2,4 dichlorophenoxy_ acetic acid (a synthetic auxin) stimulated increase in invertase activity in chicory root tissues does not involve its *de novo* synthesis. *In vitro* addition of auxin also activates membrane bound ATPase activity, although it inhibits phosphatide phosphatase activity.

Auxin effects, which have no lag period, are possibly mediated through direct enzyme regulation. A. Häger and his associates (1971) have suggested that auxin induced cell wall growth is due to plasma membrane bound ion pump which caused extrusion of H^+ from the cytoplasm into the cell wall. As most ion pumps are ATPase, increase in this enzyme activity, will cause H^+ excretion and growth. The ion pump is able to transport H^+ only when it is combined with the auxin. The effects of auxins on H^+ transport and growth is enhanced by Mg^{2+} and K^+, which might be expected to provide an ion exchange system for the H^+. The energy to drive the H^+ transport is provided by ATP generated during respiratory metabolism. The H^+ once pumped into the cell wall, would break acid labile bonds in the wall and ultimately would lead into cell wall extensiblity and growth. However, it is likely that the auxin induced cell growth involves the modulation of other enzymes as well.

Membrane Permeability: An important effect of auxins on excised plant tissues is the osmotic uptake of water causing swelling of the cell wall. It is speculated that this is caused by auxin stimulated permeability of the cell wall. It has also been suggested that auxin molecules stick to the lipid components of the membranes and alter the membrane permeability.

The possibility of altered membrane permeability as a mechanism of auxin effect is realised in fast responses of plant tissues to the hormone. Careful kinetic experiments with oat coleoptile, have shown that auxins have a stimulatory effect on growth very rapidly upon entry and that the stimulation decays very rapidly after the auxin is removed. Since enzyme synthesis should take a long time (at least more than 10-15 minutes) and since the expected half life of enzyme thus formed or activated by physico-chemical modulation, should be of at least several hours, it seems likely that the short time (2-3 minutes) and reversible effects of auxins are mediated via their effects on membrane permeability.

Involvement of a Second Messenger: The molecular mechanism of the animal hormones is much better understood that the plant hormones. In this perspective, attempts have been made to look for those details of molecular mechanisms of plant hormone action, which exist for animal hormones. It is considered that most animal hormones bind to a specific receptor protein in the cell membrane and that many of them act through a second messenger (first messenger is hormone itself) molecule, which is often Ca^{2+} and/or cyclic AMP.

Several binding sites for IAA in maize coleoptiles have been demonstrated. Some of these sites have been isolated and purified using ion exchange and gel permeation chromatography. These are proteins of molecular weight 40-45 kDa and they do not possess ATPase activity. Following its binding with a receptor protein, the auxin may activate a second messenger molecule. A.W. Galston (1983) has suggested that polyamines can act as second messenger for auxins and other plant hormones.

Polyamines are reported to increase in plants when stimulatory concentrations of plant hormones are applied. Either Ca^{2+} or the Ca binding protein, Calmodulin, can also act as second messenger. Inhibitors of calmodulin action retard the auxin induced elongation of maize and oat coleoptiles. D.C. Eliott and others (1983) have shown that there are two types of calmodulin inhibitors. At low concentrations, these inhibitors stimulate and at high concentrations prevent the morphological responses in plants due to auxins, gibberellins and cytokinins. Second messenger molecules, such as Ca^{2+}, calmodulin and polyamines, most likely act through the activation of specific enzyme activities.

Auxin Metabolism: In vascular plants, the auxins are synthesized in meristems and enlarging tissues of the shoot apex. Embryos of certain fruits and seeds also produce large amounts of auxins.

Auxin Biosynthesis: Indole acetic acid, the principal naturally occurring auxin, is synthesized from the aromatic amino acid tryptophan. Enzyme system, capable of converting tryptophan to IAA, has been isolated from spinach leaves and oat coleoptiles. IAA can also be synthesized from tryptomine and indole pyruvic acid by leaf discs or crude leaf extracts from pine apple leaves. It has been suggested that tryptophan is first converted to indole acetaldehyde either through indole pyruvic acid or through tryptomine. Indole acetaldehyde is then converted to indole acetic acid. The reactions involved are:

(1) Conversion of tryptophan to indole pyruvic acid is an oxidative transamination reaction and is catalysed by the enzyme tryptophan transaminase. Ammonia of tryptophan is accepted by a suitable keto acid, which in turn is converted to an amino acid.

(2) Conversion of indole pyruvic acid to indole acetaldehyde involves decarboxylation (of side chain) reaction and is catalysed by the enzyme indole pyruvate decarboxylase.

(3) Conversion of tryptophan to tryptomine is a decarboxylation (of side chain) reaction and is catalysed by the enzyme tryptophan decarboxylase.

(4) Conversion of tryptomine to indole acetaldehyde is oxidative deamination process catalysed by the enzyme amine oxidase.

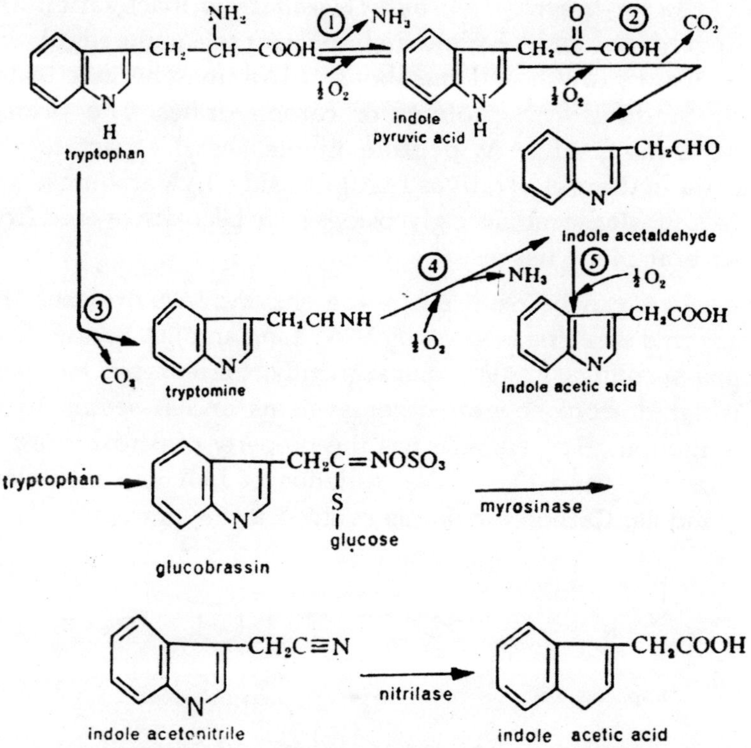

tryptophan → (1) $-NH_3$, $+\frac{1}{2}O_2$ → indole pyruvic acid → (2) $-CO_2$, $+\frac{1}{2}O_2$ → indole acetaldehyde (CH_2CHO)

tryptophan → (3) $-CO_2$ → tryptomine ($CH_2CH\,NH$) → (4) $-NH_3$, $+\frac{1}{2}O_2$ → indole acetaldehyde → (5) $+\frac{1}{2}O_2$ → indole acetic acid (CH_2COOH)

tryptophan → glucobrassin ($CH_2C{=}NOSO_3$, S, glucose) → myrosinase →

indole acetonitrile ($CH_2C{\equiv}N$) → nitrilase → indole acetic acid (CH_2COOH)

(5) Oxidation of indole acetaldehyde to indole acetic acid is the final step in IAA synthesis from tryptophan. It is catalysed by the enzyme indole acetaldehyde oxidase.

In some plants, operation of another pathway of IAA synthesis from tryptophan has been suggested. Tryptophan is converted to a glucose conjugate, glucobrassin. The enzyme for this reaction, however, has not been detected.

However, the glucobrassin has been isolated from several plants of family cruciferae and labelled tryptophan is incorporated into this compound in some plants. Glucobrassin is converted to indole acetonitrile by the enzyme myrosinase. Indoleacetonitrile in turn is converted to indole acetic acid by the enzyme nitrilase.

Auxin Inactivation and Degradation: Inactivation and degradation of IAA plays an important role in the regulation of plant growth and metabolism. IAA may be inactivated by binding with proteins or carbohydrates. The protein derivatives of IAA, peptide indole acetyl aspartate and carbohydrate derivatives IAA glucoside, IAA arabinose and IAA inositol-arabinose glycosides have been discovered from several plant tissues.

IAA can be oxidised by the enzyme IAA oxidase. This enzyme was first isolated by Y.W. Tang and J.D. Bonner from pea seedlings in 1947. Subsequently, the enzyme has been isolated from several other systems and it seems to be ubiquitous. The enzyme has the property of peroxidase and uses H_2O_2 and O_2 for the oxidation of IAA to 3-methylene oxindole. Carbon dioxide is evolved in the process;

$$\text{Indole Acetic Acid} + H_2O_2 + O_2 \xrightarrow{\text{IAA oxidase}} \text{3-methylene oxindole} + CO_2 + H_2O$$

Indole Acetic Acid 3-methylene oxindole

Several other oxidation products of IAA, such as indole acetaldehyde, 3-hydroxymethyl oxindole and 3-methyl dioxyindole have also been detected.

IAA oxidase is an extensively studied enzyme from the plant tissues. It has a Km value of 0.1 mM for IAA and requires two cofactors, a phenolic and Mn^{2+}. The phenolics which act as cofactor also inhibit plant growth as they can enhance the oxidation of IAA. The activity of IAA oxidase has an inverse correlation with growth of the tissue in many cases.

4

PLANTS AND PHYTOCHROME

Phytochrome is a photoreceptor molecule which mediates several developmental and morphogenetic responses of plants to light. The existence of this molecule was indicated by a group of plant physiologists working at the United States Department of Agriculture, Beltsville, Maryland. These scientists were interested in elucidating the mechanism of light effects of flowering. In 1952, H.A. Borthwick, S.B. Hendricks and their colleagues, postulated that a pigment received light and that it existed in two forms: active form and inactive form. The pigment was called 'Phytochrome' (abbreviated P). Phytochrome red form (Pr) was inactive and phytochrome far-red (Pfr) was active form. When Pr was irradiated with red light it was converted to Pfr. Similarly Pfr form could be converted to Pr form by irradiation with red light;

$$\underset{\text{Inactive}}{\text{Pr}} \underset{\text{Far-red (730 nm)}}{\overset{\text{Red(660)}}{\rightleftarrows}} \underset{\text{Active}}{\text{Pfr}}$$

The active Pfr form of the phytochrome could trigger developmental and morphogenetic responses in plants. This was in line with the observations that red light (which will produce Pfr) induced photomorphogenetic responses while far-red light reversed the red light induced responses. Since the effects of red or far-red light were reversible and very fast, it was concluded that a single phototransforming pigment was involved in the effect of light.

The pigment phytochrome, was ultimately isolated by W.L. Butler and his colleagues (1959, 64) also from USDA, Beltsville. It was recognised as the first protein, other than enzymes, to have regulatory functions in plants.

Significant Aspects

Isolation of Phytochrome

Phytochrome is a protein and it can be extracted and isolated from seedlings using standard procedures of protein isolation. The pigment is precipitated out from the extract by ammonium sulfate. Then it is purified using column, ion exchange and gel exclusion chromatography. However, these methods are time consuming and yield poor quality of phytochrome.

More refined techniques have been used to obtain good quantity and quality of phytochrome, one of the major problems in the conventional methods was degradation of phytochrome by endogenous proteolytic activity.

Polyethylene amine was used to precipitate out acidic proteins, nucleic acids, and photosynthetic pigments from crude extracts.

Affinity chromatography utilising antiphytochrome aminoglobulins has been used to obtain intact phytochrome molecule. Phytochrome isolated from a variety of sources including oats, rye, zucchini, pea, lettuce, etc., is first partially purified by brushite chromatography and ammonium sulfate fractionation. The immunoglobulins for these preparations are raised, and are immobilized in a column. The mixture containing phytochrome is allowed to run through this column. While phytochrome is retained in the column by binding with its specific immunoglobulins other proteins flow down the column.

The phytochrome is then eluted from the column by using 3 M $MgCl_2$ at about pH 7.5. Subsequent elution with 1 N formic acid gives additional yield of phytochrome.

Affinity chromatography of phytochrome using agarose immobilized Cibacron Blue F_3 has also been used successfully to purify phytochrome. In this, the pigment is eluted with a buffer containing 10 mM flavin mononucleotide (FMN).

The quantity of phytochrome isolated is determined by using spectrophotometric methods. Immunochemical and radio-immunoassay methods also have been developed for its measurement.

Distribution of Phytochromes

Phytochrome has been detected in a wide variety of plants. In addition to, having been isolated from such higher plants as tobacco, oat, maize, bean and rye, etc., phytochrome has also been isolated from the alga *Mestaenium* and the liverwort *Sphaerocarpos.* In higher plants, it has been detected in roots, stems, hypocotyls, cotyledons, coleoptiles, leaf blades, petioles, vegetative buds, developing fruits, floral receptacles and inflorescences.

As far as the cellular distribution is concerned, phytochrome is thought to be associated with etioplasts, chloroplast membranes are plasmalemma. Some scientists suggest that the action of Pfr is dependent on its association with membranes and alteration of membranes.

Physico-Chemical Nature

Two types of phytochromes; small and large, have been identified. The small phytochrome has a m.w. of about 60 kDa. It is believed to be degradation product of large phytochrome and does not appear to exist *in vivo.* Large phytochrome appears to be the natural phytochrome.

It is a dimer with a m.w. of each monomer as 120kDa. Recent evidences indicate that even this large phytochrome is 5% smaller than the native molecules, which appears to be a

124kDa polypeptide. Apparently proteolytic activity degrades it during extraction and isolation, to yield a 120 kDa polypeptide.

Phytochrome is a conjugated protein, since it contains a visible chromophore. Because of this chromophore, it appears bluish. The protein component is a water soluble globular protein. The molecule is a dimer with two chromophores and two globular proteins. There are three disulfide (S—S) bonds per monomer, but these bonds are not involved in holding the sub units together.

The two monomeric units are identical, each made up of about 1100 amino acids. About 46% of the amino acids are polar (lysine, histidine, arginine, asparagine, serine, glutamate and glutamine). The protein portion to which chromophore is attached contains the sequence; leucine — arginine — alanine — proline — histidine — serine— cysteine — histidine — leucine — glutamine — tyrosine. The chromophore is linked through cysteine residue.

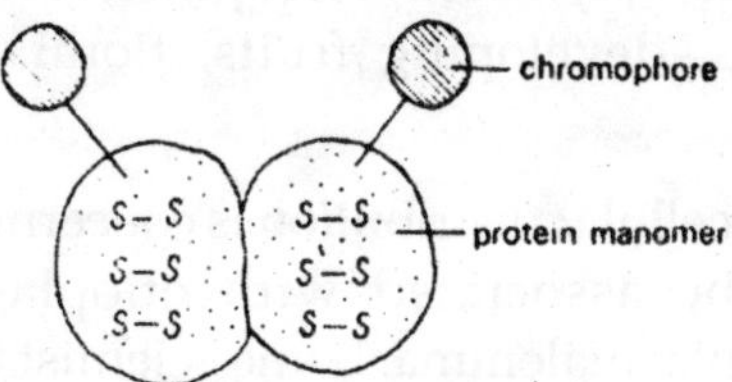

Diagrammatic representation of a phytochrome molecule.

As far as the secondary structure of the phytochrome protein is concerned, nearly half of the protein is composed of equal amounts of alpha-helix and beta structure, while remainder is aperiodic.

Chromophore

Although the exact number of chromophore molecules is not known, in all probability there is one chromophore per monomeric unit. The basic structure of chromophore is essentially similar to the algal pigment 6-phycocyanin. Binding of the chromophore to the protein involves a thio-easter linkage through cysteine to the C-2 side chain of ring A. Perhaps the B ring is also involved in the binding with the protein, through its side chain $(CH_2)_2COOH$. When phytochrome undergoes Pr ⇄ Pfr interconversions, the configuration of A unit is changed. In addition to the chromophore, the phytochrome also contains about one phosphate per monomer. The function of phosphate, is however, unknown.

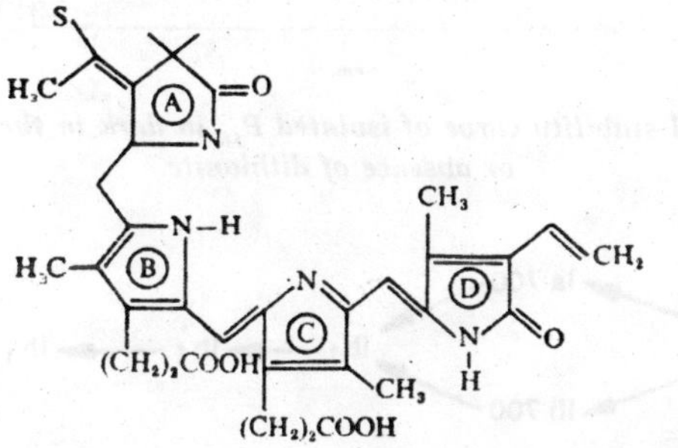

The chromophore moiety of the phytochorme.

Pr ⇄ Pfr Conversions: As mentioned earlier in this chapter, red light converts phytochrome red (Pr) in to phytochrome far-red (Pfr) and far-red light reverses this process. The Pfr is converted to Pr form in dark as well, although the process is slower. The conversion of Pfr to Pr in dark is accelerated by

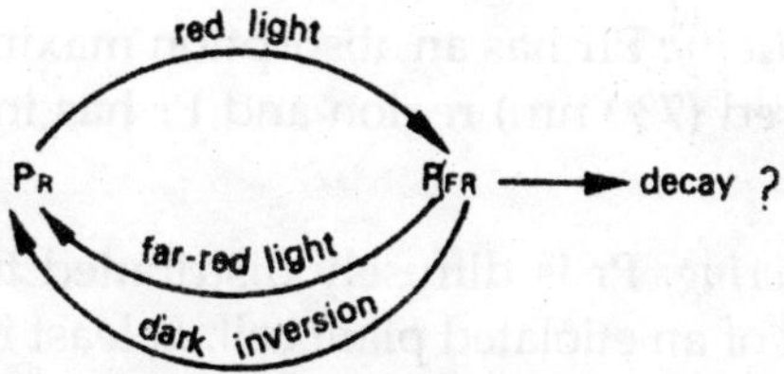

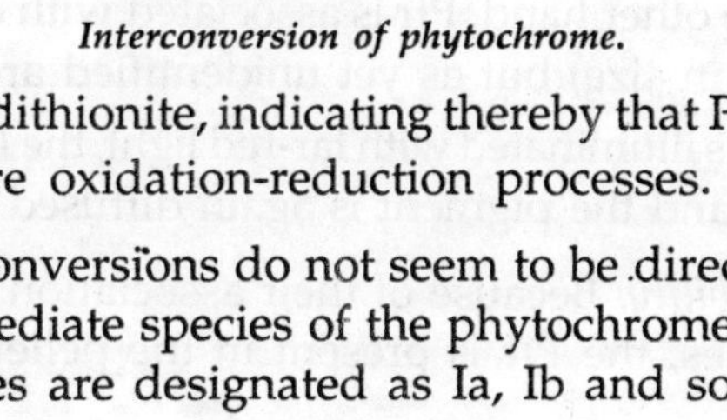

Interconversion of phytochrome.

the reductant dithionite, indicating thereby that Pr ⇄ Pfr interconversions are oxidation-reduction processes.

The interconversions do not seem to be direct but involve several intermediate species of the phytochrome. These intermediate species are designated as Ia, Ib and so on.

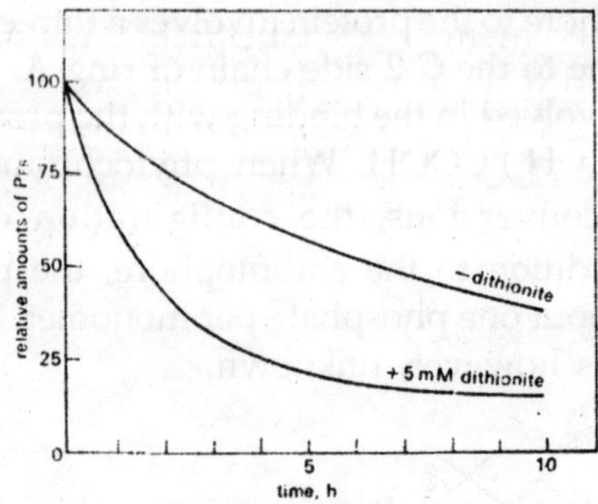

A typical stability curve of isolated P_{FR} in dark in the presence or absence of dithionite.

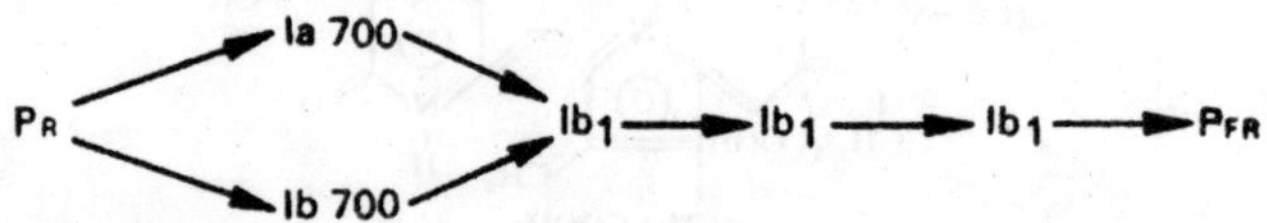

Some intermediates in the conversion of P_R to P_{FR}

The two species of phytochrome Pr and Pfr, differ from each other in their following properties:

1. *Activity:* Pfr form of the phytochrome is active form while Pr is inactive. That is to say, a phytochrome mediated response is seen only when it is in Pfr form.
2. *Absorptivity:* Pfr has an absorption maximum (λ , max) in far-red (730 nm) region and Pr has in red (680 nm) region.
3. *Sequestering:* Pr is diffusely distributed throughout the cytosol of an etiolated plant cell, at least in some plants. On the other hand, Pfr is associated with discrete (about 1 μm in size) but as yet unidentified areas. When the tissue is illuminated with far-red light, the Pfr is converted to Pr, and the pigment is again diffused in the cytosol.
4. *Pelletability:* Because of their association in the form of particles, the Pfr is present in the pellets, while Pr in

the supernatant of a plant extract. This fractionation is achieved when centrifugation of the extract is done at 20000 xg. Following conversion of Pr to Pfr, the amount of phytochrome to be obtained in pellets increases several folds.

5. *Reactivity to Reagents:* Pfr is more active towards urea, metal ions such as Cu^{2+}, Co^{2+} and Zn^{2+} and N-ethyl maleimide, than Pr.

These differences in the properties of Pr and Pfr are possibly because of reorientation of the phytochrome molecule during interconversions. Besides, reduction-oxidation of chromophore moiety, some changes in the organization of protein component also seem to take place.

Phytochrome Mediated Responses

Seed Germination: Germination of many types of seeds is enhanced by light. Phytochrome has been shown to be involved in this response of light, as in most cases red light enhances germination and far-red light inhibits it. The way in which the active phytochrome brings about its effect on germination is not clear, but some intermediate metabolites seem to be involved in the process.

Seedling Growth: The effect of light on several aspects of seedling and plant growth seem to be mediated via phytochrome. The growth of coleoptile, the mesocotyl extension, unrolling of grass leaves, etc. are a few examples of phytochrome mediated processes.

Growth of mature plant is affected by light through phytochrome. If a dicot plant or a conifer is grown in far-red light (as in shaded conditions), its stem becomes considerably elongated. Branching in many species is also retarded. The effects of far-red light or shading are apparently caused by decreasing the level of Pfr compared to that obtained in unshaded plants.

Flowering: The photoperiodic flowering responses in most species has been shown to be mediated via phytochrome. In

Xanthium for example, where flowering is inhibited by exposure to light during desired dark period, the far-red is without effect. These plants behave as though they were exposed to continuous darkness. Even a small flash of far red light, negates the inhibitory effect of red light and flowering takes place.

The precise role of phytochrome in flower initiation is not known. However, it is believed that the pigment is involved in triggering the formation (synthesis or activation) of the flowering stimulus. The level of phytochrome Pfr, the physiologically active form, depends upon a number of factors.

Flavonoid Biosynthesis: Flavonoids impart characteristic colour to the plant parts. Their synthesis is induced by both low and high energy radiations. For the low energy radiations, phytochrome has been shown to be the exclusive pigment receptor. In high energy reactions, however, pigments other than phytochrome, such as carotenoids, flavins and cytochromes, are also involved. The plant species where phytochrome mediated flavonoid biosynthesis has been observed include *Amaranthus*, barley, buckwheat, red cabbage, maize, *Sorghum*, turnip, etc.

Physiological and Biochemical Responses: Many effects of phytochrome at morphological and morphogenetic level arise due to changes in the physiology and biochemistry of the plant. Some physiological and biochemical changes are independent of any direct morphological effect, as well. Some of the physiological changes are: change in the rate of cellular respiration, membrane permeability, change in auxin catabolism and in cell turgidity. Some biochemical changes include increase in RNA and protein synthesis and changes in some enzyme activities. The enzymes of CO_2 assimilation and starch biosynthesis, which are located in chloroplasts, are activated by light and deactivated by dark. The amylase activity in *Sinapis alba* the phenyl-alanine ammonia lyase activity in cultured parsley cells, and nitrate reductase activity in a few plants is activated by Phytochrome.

Other Facets

The Mechanism

The phytochrome mediated responses appear to be of two types: (i) Fast responses, which are seen within a few minutes of desired irradiation, such as change in membrane permeability and in cell turgor and (ii) slow responses, which are seen after hours, days or weeks after irradiation, such as bulb formation and flowering.

In both types of responses, it appears that the primary target of phytochrome action is the membrane. It is however not clear, whether the plasma membrane or the organellar membrane is affected when phytochrome is converted to its active form (Pfr). However, it is clear that permeability of the cell to water and ions such as H^+, K^+ and Ca^{2+} is changed. G. Robin and others (1989) have suggested that the phytochrome mediated pulvinar movement of *Cassia fasciculata* involved Ca^{2+} and Ca^{2+}-calmodulin action. Perhaps Pfr promotes movement of Ca^{2+} from outside to the cytoplasm. This Ca after complexing with calmodulin changes the permeability of membranes for ions such as H^+, K^+, Cl^-, etc.

The change in membrane permeability may be manifested in a variety of physiological and biochemical responses. Changes in the membrane permeability may be associated with appearance of a specific charge on the plant surfaces, as is observed in the root tip. During Pr $\rightleftarrows$ Pfr interconversions, some cations such as H^+ are preferentially held at outer surface of the root. It results in a positive charge on the root surface. This can be demonstrated in attraction and attachment of these root tips on a negatively charged glass surface. This phenomenon has been termed as *Tanada effect*, as it was first discovered by Takuma Tanada at Beltsville, U.S.A.

If phytochrome is converted from Pfr to Pr, the charge on the root surface is neutralized, and the root tips are readily released from the glass surface. At biochemical levels, changes in the cellular concentration of H^+, K^+, Ca^{2+} and other solutes may modify enzyme activities, which may set in a motion a chain of biochemical reactions which may lead to various morphogenetic responses.

It is also likely that the conversion of Pr to Pfr activates specific genes and the eventual synthesis of new enzymes. Researches conducted by Hans Mohr and others in Germany in 1960's and 1970's lead them to hypothesize that Pfr somehow activates previously repressed genes, so that the m-RNA and ultimately the specific enzymes could be synthesized. This hypothesis was supported by the facts that in many cases inhibitors of m-RNA formation (transcription) and its translation such as actinomycin-D and puromycin prevent phytochrome mediated increase in enzymes. *De novo* synthesis of enzymes has also been demonstrated by using labelled amino acids and their incorporation into proteins. In all these responses there is a lag of about 2-3 hours and thus the hypothesis of gene mediated enzyme synthesis can explain only these slow responses of phytochrome.

High Irradiance Responses

Phytochrome mediated responses are evoked by relatively lower levels (intensity and duration) of light and are therefore called low irradiance responses. That is to say even a short exposure to low or moderate light intensity can bring about those effects of light which are mediated via phytochrome. However, several other photomorphogenetic responses proceed only when a long exposure to low or moderate light or a short exposure to high intensity light is given.

Such responses are called as *High Energy Reactions (HER)* or *High Irradiance Responses* (HIR). An example of HIR is cotyledon enlargement in *Sinapis alba*. The enlargement is induced by blue light (450 nm) and far-red (730 nm) light but not by red (680 nm) light. The response to blue light is a characteristic of HIR. Some other HIR are evoked by near ultra violet light of about 370 nm.

The major distinction between HIR and phytochrome mediated responses is that phytochrome mediated responses are red-far-red reversible, whereas HIR are not. It is not clear whether some special form of phytochrome or some other pigment is involved in HIR.

The Ethylene

Although, the interest in ethylene developed primarily as its pollutant nature, researches conducted mostly in the later half of this century have clearly demonstrated that it has growth regulating properties as well. It is produced by almost all parts of higher plants, and thus it is not only a gaseous hydrocarbon but a plant metabolite as well. Like abscisic acid, ethylene is also usually considered as an inhibitory hormone.

The history of the discovery of ethylene as a plant hormone, starts with 19th century when plant growers noticed the effect of smoke and burning gases on plants. But the first recorded effect can be credited to the Russian botanist Dimitry N. Neljobow (1901) working in St. Petersburg, who demonstrated that ethylene affects plant growth. He also established that ethylene was emanated from illuminating gases and it had three types of effects on pea seedlings: stem elongation, stem thickening and horizontal growth. Many other plant physiologists also demonstrated the effect of illuminating gases on plants. In 1912, A.F. Sievers and R.H. True demonstrated that the combustion gases from the kerosene stove were beneficial in the ripening of lemons. W.C. Crocker and L.I. Knight (1908, 1913) identified ethylene as the active constituent of both, illuminating gas and tobacco smoke. E.M. Harvey (1913) determined that biological responses produced from burning paper could be attributed to ethylene. In 1942, F.E. Denny confirmed that ethylene was active substance in kerosene fumes causing the ripening of lemons.

The first report that plant materials evolve ethylene came from H.H. Cousins in 1910, who observed that when oranges and banana were stored together during shipping, some gas was emited from oranges which caused ripening of banana. In 1932, O.H. Elmer reported that the sprouting of potatoes could be inhibited by keeping them in close proximity of mature apples or pears. In 1934, R. Gane provided chemical proof that ethylene was produced by ripe apples. Since then, it has been shown that ethylene is produced from essentially all parts of higher plants.

Ethylene is introduced into the atmosphere from both physical as well as vegetable sources. In coal gas, its concentration is in order of 10 ppm and in burning stubbles upto 4 ppm. In car exhausts it may be as high as 400 Ppm. In most urban areas, its mean concentration does not normally exceed 0-50 ppb and in rural areas it may be less than 5 ppb.

Bioassay

Because of its gaseous nature, ethylene is generally detected by gas chromatography. It can also be measured by manometric method. The gas is absorbed by mercuric perchlorate solution and liberated again by chlorides. For this, the gas or the test vapours are bubbled through a small volume of mercury perchlorate or through silica gel impregnated with mercury perchlorate. Then HCl or LiCl is added to it, which causes the liberation of ethylene gas. The volume of the gas can be measured and quantified. This procedure can be used for measuring the amount of ethylene as well as for concentrating it.

Detection of ethylene using plant materials as tests, is also used in some investigations. The triple response of etiolated pea plant as developed by Neljubow (1901) is still used as bioassay. In most cases the plant material to be assayed for ethylene is enclosed in a small chamber, with an injection port. After incubation of weighed amounts of the tissue for required time, samples of gas are withdrawn with a hypodermic syringe and tested on other plant. Following bioassays are used:

1. *Reduction in Extension Growth:* Plants enclosed in a chamber containing 0.01 ppm ethylene show reduction in extension growth. While at 1.0 ppm there is increase in horizontal growth and stem swelling takes place. Besides etiolated pea seedlings, sunflower, tomato,

potato, buckwheat and castor bean have also been used for this type of bioassay.

2. *Acceleration of Senescence:* Ethylene accelerates senescence of leaves, flowers, petals, etc. When leaves are enclosed in a chamber containing ethylene, loss in chlorophyll content can be recorded. Senescence of cotton cotyledons is also accelerated, and this is used as bioassay.

3. *Induction of Stem and Leaf Epinasty:* Epinastic responses in tomato plants is also used sometimes as ethylene bioassay.

Physiological Effects

1. *Seed Dermination:* Ethylene is known to break dormancy and induce germination of lettuce, ground nut, wheat, clover, and cocklebur seeds. It also causes the increased extension growth of the seedling in cocklebur. The extension growth is maximum at 0.3 ppm ethylene. The maximum germination however, is obtained at about 40-50 ppm ethylene. On the other hand, inhibition of germination by ethylene has been recorded in maize, *Potentilla* and some weeds.

2. *Growth Inhibition and Morphogentic Effects:* In most cases exogenous application of ethylene inhibits plant growth. In most dicots the elongation growth of stems, roots and leaves is inhibited. But the hormone enhances radial growth, as a result both stem and root swell in response to ethylene. Auxin and ethylene when applied exogenously or produced in partially inhibitory concentrations, act synergistically to inhibit root elongation and swelling of the root tips.

Growth of roots, when measured in terms of dry weight is unaffected by 10 ppm ethylene for 11 days in barley, although the same treatment causes stunting and feathery growth of the roots. Similar concentration of ethylene however, inhibits dry weight in tomato.

Reduction in growth may be due to inhibition of cell division, as has been observed in various tissues of etiolated pea plants. In most cases, inhibition of growth is proportional to the inhibition of cell division. In pea seedlings, ethylene inhibits xylem vessel formation also. But no permanent damage is done to the cells; as soon as ethylene is removed, normal differentiation takes place.

Some growth promoting effects of ethylene are also known. The hormone appears to be involved in the increased coleoptile and shoot growth of submerged plants such as rice. Growth of rice coleoptile is greatly accelerated when rice plants are submerged in water. This is believed to be due to the combined influence of decreased oxygen supply and the accumulation of growth promoting concentration of ethylene and CO_2.

Each component appears to be acting independently. In rice plants growing in air, supply of ethylene greatly accelerates the growth of internodes.

One of the most important effects of ethylene is on hook opening. It inhibits hook opening in etiolated lettuce and bean seedlings.

3. *Epinastic Responses:* Exposure to ethylene causes epinastic movement in petioles, as a result the leaves bend down. This is because of more growth on upper side than on lower side of the petiole. Epinasty is also caused when excess of auxin is applied. This is believed to be due to endogenous ethylene production. Stem epinasty is also caused due to ethylene, as in tomato.

4. *Flowering Inhibition and Sex Expression:* Ethylene inhibits flowering in most plants, although it is known to promote flowering in mango and pine apple. In *Plumbago indica* also, a short day plant, flowering can be induced by ethylene in long days. Flowering induction in pine apple with response to certain auxins has been attributed to endogenous ethylene production.

Inhibition, of flowering by ethylene is controlled by the photoperiod. In *Pharbitis nil*, the hormone inhibits flowering only when it is applied during inductive dark period, and not when it is applied before or after inductive dark period. It is suggested that ethylene applied during the dark period produces some flower inhibiting substance.

Ethylene changes sex expression also in unisexual plants. It increases female flowers in several members of cucurbitaceae. It also induces male sterility in cucurbits and wheat.

5. *Ripening of Fruits:* Acceleration of fruit ripening was the first discovered effect of ethylene. The hormone is now known to accelerate ripening of mature fruits in most cases including banana, apple, tomato avocado, etc. In apple fruits, the softening of fruits increases with the increase in ethylene concentration, reaching maximum at 50 ppm ethylene. In most fruits, this softening is brought about by chemical changes in the cell was materials.

   Ethylene appears to play an important role in the natural ripening of fruits. During ripening of most fruits, the ethylene production increases sometimes by as much as 50 to 100 folds. The ripening process of fruits can be delayed by removing ethylene from the vicinity of developing fruits or by maintaining low oxygen levels around fruits.

6. *Acceleration of Senescence and Abscission:* Ethylene is an important hormone governing the senescence and abscission of plant parts, both natural and induced. It accelerates senescence of leaves, flowers and fruits. Endogenous ethylene increases during senescence. Supply of ethylene degrades chlorophyll. Aminocyclopropane carboxylic acid (ACC), a precursor in ethylene biosynthesis, also accelerates chlorophyll degradation in detached leaves.

Abscission is the most widely demonstrated response of ethylene. It induces abscission of leaves, fruits, petals and flowers (Table). The threshold concentration to induce abscission is generally 1 ppm, although in some cases as low as 0.01 ppm ethylene (as in fox glove corola) can induce it.

The abscission increases with ethylene concentration, saturating at about 10 ppm- Exposing fruiting cotton plants to 0.5 ppm ethylene causes 100% abscission of young fruits and floral buds within 2 days. In this respect, it is believed that the initial effect of ethylene is to lower the auxin content.

Some Examples of Abscission Systems where Ethylene or Ethylene Generating Chemicals have been Shown to Accelerate Abscission.

Leaves	*Young fruit*	*Petals*	*Flowers*
Pecan	Avocado	Geranium	Begonia
Bean	Apple	Foxglove	Tobacco
Olive	Orange		Apple
Orange	Pecan		Ecballium
Mung bean	Prune		Pistachio
Cotton	Mature fruit		Zygocanctus
Sambucus	Orange		Lilium.
Coleus	Tomato		Leptospermum
Rubber	Peppers		Tomato
Mulberry	Grape		
Poplar	Olive		
Eucalyptus	Lemon		
	Cherry		
	Ecballium		
	Coffea		

Mechanism of Action

1. *Involvement of a Receptor:* Like most other plant and animal hormones, the action of ethylene is initiated by

its binding to a specific receptor site in the target cell. The existence of a receptor site has been demonstrated in various tissues and organs of a variety of plant species. The site is a membrane bound protein perhaps an enzyme containing Cu atom. The ethylene binds to the Cu. The same receptor also recognises CO and CO_2. Carbon mono-oxide causes some of the same physiological effects as ethylene. But CO_2 is competitive inhibitor of ethylene action.

2. *Increased Membrane Permeability:* Ethylene is very soluble in lipids and thus could easily penetrate into cell membranes. Exogenously supplied ethylene has been shown to increase permeability in a variety of plants. In barley aleurone layers, it accelerates the secretion of a-amylase, but not its production.

   The hormone also causes the swelling of isolated mitochondria, which is taken as an indication of increased permeability. The increased permeability in senescing petals of *Tradescantia* involves increased RNA and protein synthesis. It has been suggested that ethylene induces phospholipases which degrade phospholipids and cause leakiness of the membrane.

3. *Activation of Gene Expression:* It is believed that at least some responses of ethylene such as fruit ripening and abscission are due to the activation of gene expression and increased protein synthesis.

   The nucleic acid and protein synthesis often increase during increased ethylene production in ripening of fruits. Several enzymes such as polygalacturonases, a cell wall softening enzyme, increase due to their activated synthesis. Other enzymes such as phenylalanine ammonia lyase, cellulase, polyphenol oxidase, peroxidase, etc. also increase in response to ethylene.

   The increase in phenylalanine ammonia lyase is clearly due to its *de novo* synthesis. This enzyme catalyses the deamination of 1-phenylalanine to transcinnamic acid, which can then be incorporated into numerous phenolic compounds including anthocyanidins. The anthocyanidins impart bright colour to the fruits and their increased production is an important aspect of fruit ripening. Increased cellulase and some related enzyme activities may be the factor involved in softening of the fruits during ripening.

4. *Involvement of Auxins:* Auxins induce ethylene production and many responses of ethylene are similar to that of auxins. These include epinasty, inhibition of stem, root and leaf elongation, flower induction in pine apple and mangoes, inhibition of epicotyl or hypocotyl, hook opening in dicot seedlings, etc. It is likely that mechanism of action of both hormones is either the same or similar.

Ethylene Metabolism

Biosynthesis: All higher plants produce ethylene. By feeding labelled metabolites it has been shown that the primary precursor of ethylene is the amino acid, methionine. Third and fourth carbon (C-3 and C-4) of methionine is converted to ethylene.

Other eventual products of methionine metabolism are CO_2, formic acid, ammonia, and methyl S-adenosine or some other methyl compound. Both the ammonia and methyl thio group can be metabolised back into methionine thereby conserving nitrogen and sulfur and allowing ethylene synthesis to continue.

$$\begin{array}{c} \qquad\qquad\qquad\qquad\qquad\qquad NH_3 \\ \qquad\qquad\qquad\qquad\qquad NH_2 \\ {}^5CH_3-S-{}^4CH_2-{}^3CH_2-{}^2CH_2-{}^1COOH \end{array}$$

methyl pool ethylene HCOOH CO_2

Origin of ethylene and other metabolites from methionine.

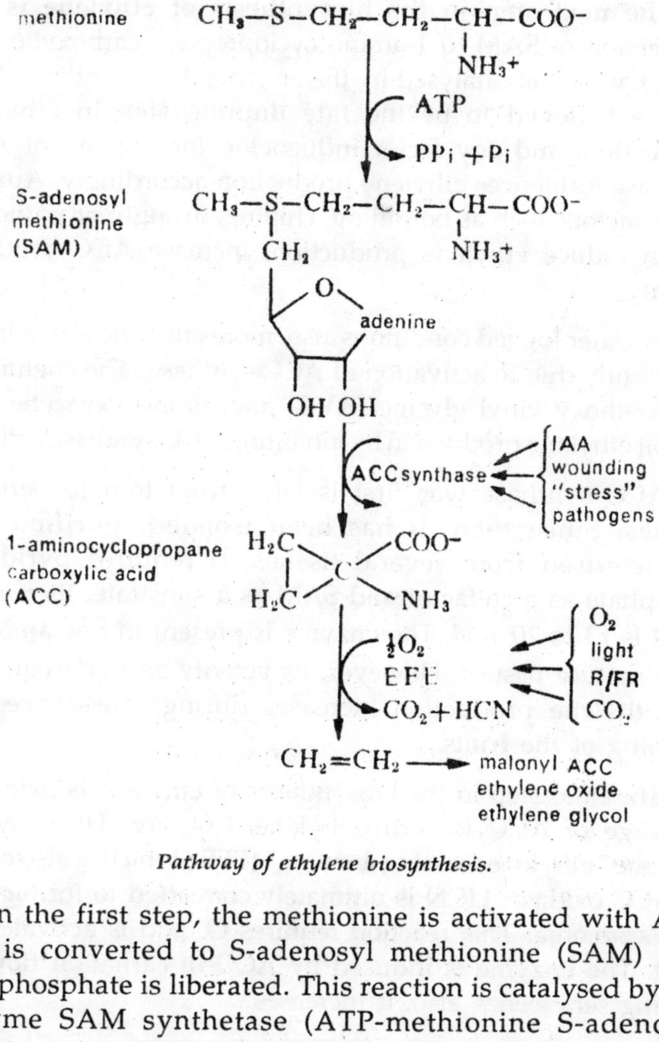

methionine $CH_3-S-CH_2-CH_2-CH(NH_3^+)-COO^-$

ATP → PPi + Pi

S-adenosyl methionine (SAM) $CH_3-S^{(CH_2\text{-ribose-adenine})}-CH_2-CH_2-CH(NH_3^+)-COO^-$

O, adenine, OH OH

ACC synthase ← IAA, wounding, "stress", pathogens

1-aminocyclopropane carboxylic acid (ACC) H_2C, H_2C, C, COO^-, NH_3

$\frac{1}{2}O_2$, EFE → CO_2 + HCN ← O_2, light, R/FR, CO_2

$CH_2{=}CH_2$ → malonyl ACC, ethylene oxide, ethylene glycol

Pathway of ethylene biosynthesis.

In the first step, the methionine is activated with ATP and is converted to S-adenosyl methionine (SAM) and pyrophosphate is liberated. This reaction is catalysed by the enzyme SAM synthetase (ATP-methionine S-adenosyl transferase). In SAM, the sulfur atom of methionine is linked to the C-5 of the ribose moiety in adenosine. In many cases, solenomethionine has been found to be the better substrate for ethylene biosynthesis than methionine. The enzyme SAM synthetase has a higher affinity for solenomethionine than for methionine.

The next step in the biosynthesis of ethylene is the conversion of SAM to 1-amino cyclopropane carboxylic acid (ACC), which is catalysed by the enzyme ACC synthase. This step is believed to be the rate limiting step in ethylene production, and any factor influencing the activity of ACC synthase influences ethylene production accordingly. Auxins, stress factors such as bounding, chilling, drought or pathogen which induce ethylene production, increase ACC synthase activity.

In water logged conditions also, more ethylene is produced apparently due to activation of ACC synthase. The chemicals, aminoethoxy vinyl glycine (AVG) and amino oxyacetic acid inhibit ethylene production by inhibiting ACC synthase activity.

ACC synthase was first isolated from tomato pericarp tissues. Since then, it has been isolated, purified and characterised from several tissues. It requires pyridoxal phosphate as a co-factor and SAM as a substrate. Its Km for SAM is 13 to 20 μM. The enzyme is present in low amounts in immature tissues. However, its activity and consequently the ethylene production increases during senescence and ripening of the fruits.

The final step in the biosynthesis of ethylene is oxidative cleavage of ACC to form ethylene, CO_2 and HCN by the enzyme "ethylene forming enzyme" (EFE), which is also called as ACC oxidase. HCN is ultimately converted to formic acid and ammonia. This reaction requires O_2 and is activated by light. The enzyme is induced by ACC in carnation flowers. During senescence also, it increases.

The enzyme is thought to be located in membranes, including plasma membrane and tonoplast. Its nature is largely unknown, but it is known to act through an enzymic generation of free radicals. Peroxidases and lipoxygenases have been proposed to play the role of this enzyme.

Regarding the site of ethylene biosynthesis, the shoot apex appears to be the principal site in seedlings. Nodes of dicot seedlings produce more ethylene than internodes. Roots produce relatively small amounts of ethylene. Leaves, flowers and fruits also produce ethylene, which increases during senescence and ripening.

Catabolism: Plants and micro-organisms are known to metabolise ethylene as well. The product of metabolism appear to be ethylene oxide, which is incorporated into glyoxylate cycle as ethylene glycol and malonyl ACC. Ethylene may also form a conjugate with glucose.

Commercial uses of Ethylene

Ethylene has been used for synchronised flowering and fruit ripening for centuries. Since nineteenth century, PuertoRican and Philippine pine apple growers lit bonfire near their plantation, knowing that smoke helped to initiate and synchronise the flowering of their crops and ripening of fruits. Several chemicals, which release ethylene are used now-a-days for promotion of flowering in pineapple, all over the pineapple producing world.

Ethrel or Etephon is the most commonly used chemical, which is 2-chloroethyl phosphoric acid ($Cl—CH_2—CH_2—PO_3H_2$). It rapidly breaks down in water at neutral or alkaline pH to produce ethylene, Cl^- and PO_4^-. This and some other uses of ethylene which are related to its physiological effects are listed in Table.

Possible Application of Ethylene Related to its Physiological Effects

Physiological effect	*Possible use*
Growth Inhibition	Prevention of lodging, mainly in cereals.
Stimulation of germination	Improvement of weed control in some weed species with long lasting germination periods.
Induction of female flowers in cucurbitaceae	Increase in yield and earlier harvest in cucurbitaceae.

Contd....

Physiological effect	*Possible use*
Promotion of flowering	Improved management in pine apple farms.
Promotion of ripening	Yield increase and harvest aid in tomato, harvest aid and quality improvement in tobacco, improvement of quality and colour in fruits and vegetables.

5

Controlling Insects

Insect control in broader sense may be defined as to make the life of an insect harder or to prevent their increase or spread or to kill them by any means. It depends upon the knowledge of the factors causing or favouring the abundance of the particular species. The control of insects can be achieved by several ways, which may be classified as under:

Natural Control

In this method the control is going on independently and human being have no any relation with this. The different ways are as follows:

1. Climatological factors—temperature, rainfall, light, atmospheric pressure and air currents.
2. Physical factors—mountain, seas, rivers, lakes, desert and forest etc.
3. Natural enemies—parasites, predators including insects, birds, reptiles and mammals, cannibalism etc.
4. Natural diseases—may be caused by entomogenous fungi, bacteria, virus, protozoa and other micro-organisms.

Artificial or Applied Control

It is performed by the man and this may be of following type:

(A) Chemical control-Chemicals are used to check the population of insect.

 (a) Insecticides.

  (i) Stomach poison

  (ii) Contact poison

  (iii) Fumigants

 (b) Attractants;

 (c) Repellents and Antifeedants

 (d) Pheromones and Hormones

 (e) Chemosterilants

 (f) Auxiliary, synergistic or supplementary substances

(B) Mechanical and physical control

(C) Cultural control including use of resistant varieties

(D) Biological control

(E) Legal control

Natural Control: All methods that destroy or check the infestation and which do not depend upon man for their success or continuance are collectively included in this method. The natural factors that check the insect population, their distribution and activities are as follows:

Climatic Factors

Climate is the most important natural force in keeping insect population under control. It includes temperature, rainfall, sunshine, atmospheric pressure and air currents etc.

Temperature: It is the most important factor of the environment. Besides having a direct effect on the development and metabolism of the organisms which determines their abundance, it also affects them indirectly through food and other environmental factors such as moisture, movement of air and-atmospheric pressure etc. Temperature controls the development, fecundity, feeding, movement and disposal or distribution of insects. The insects are cold blooded animal, the body temperature of which depends more or less on that of the environment. A warm, moist environment is favourable for the development of most insects. Extremes of temperature are destructive toinsect life, the optimum temperature for most of the species range between 13 to 40° C and above this aestivation begins while below this hibernation takes place.

It is clear from this that they remain active only for a particular time and in most of the months they are forced to remain inactive by nature itself due to the temperature fluctuation. The honey-bees, however, live in an environment of their own creation and are able to 'air condition' their nests.

Rainfall and Humidity: After temperature, the moisture is another important factor which affects the insects in several ways. The body composition of all insects, like other organisms, consists mostly water and they contain 80-90% water. Since water is an essential constituent of the cell protoplasm, therefore, all the processes inside the body require water. One of the fundamental necessities of all the insects is to maintain a proper balance between the moisture contents of their body and that of their environment. The activity of some insects is dependent on the advent of rainfall, *e.g.*, ants, termites, white grub, grasshoppers, hairy caterpillars etc. Rainfall influences the insects not only by the total amount of the area but also by its distribution in time. It affects the breeding, multiplication, and out break of pests, *e.g.*, the red pumpkin beetle does not lay eggs, until it comes across moist soil and general opposition takes place by this beetle only on days following irrigation or a heavy shower of rain. The adults of red hairy caterpillars (*Amsacta moori*) though fully formed inside the pupae do not emerge till there are one or two showers of rain in June or July.

Light: Sunlight is the greatest single source of energy for almost all biological systems and there are different rays *viz.* X-rays, gamma rays, ultraviolet rays and cosmic rays which directly affect the development, reproduction and growth of insects. The growth and moulting of insects are accelerated or retarded according to their requirements of light. The direct sunlight itself may kill the hidden stages of insects, therefore, they tried to hide themselves and become active in its absence. It also regulates some other activities of various insects. There is correlation between the activities of honey-bees and hours of sunshine. Most of the cock chaffer beetles and many moths pass the day in concealment. Dusk is the most usual time for the flight and copulation of moths; for the emergence of winged white ants etc. The number of hours of light in a day length (24 hrs.), termed photoperiodism, has considerable influence on the development of aphids. By reducing the photoperiod the production of sexual forms is induced. Thus each insect has its own light requirement and slight variation on this may affect their life processes.

Atmospheric Pressure: The atmospheric pressure like other factors also affect the activities of insects. The phototropic insects are more active during periods of high barometric pressure and specially when the barometer is rising. The insects are much more affected in low pressure areas than the high atmospheric pressure areas. The activities of some insects are directly influenced by the pressure for example due to low atmospheric pressure the chances of rain increased resulting the emergence of ants with their eggs for their safety.

Air Currents: The disposal of insects to a great extent depends upon the wind as they generally fly with the air current although some insects like mosquitoes habitually fly against it.

Many insect species are known to spread to new countries through air currents. Large number of aphids have been found after a strong wind and many of them are destroyed by falling in the seas and rivers etc. Thus air currents also play an important role in natural control of insects.

Physical Factors

Physical factors play important role in the distribution and abundance of insects. They include the mountains, seas, lakes and rivers, deserts, physical and chemical structure of soil and forests:

Mountain: Large mountains provide effective barriers to the insect spread as they are unable to cross them more over in high peaks, throughout the year ice is found due to which the insect can not crawl or cross them and ultimately their spread is checked.

Sea, Lakes and Rivers: Some species of insect do not possess the power of flight, hence they can not cross the seas, lakes and rivers and finally their spread is restricted.

Soil Structure: The soil type and its fertility determine the plant and animal life which in turn tend to limit the number and type of species and their abundance. Similarly the physical and chemical composition of the soil may determine the presence or absence of soil inhabiting species, for example the white grub of groundnut prefer the sandy soil and it can not thrive in clay soil.

Deserts: The insect can not cross the large deserts and thus they act as barrier. Moreover, due to the absence of vegetation and different soil texture inhibit the spread of insects.

Natural Enemies

Every insect has its natural enemies in the nature and with the increase in the population of insects, the natural enemies begin to increase as they get larger food supply. The important natural enemies are insect parasites and predators, birds, mammals and reptiles.

Insect Predators: They feed on the living insects and consume more than one individual during their life span. The important predatory insects are as follows:

Coccinellids: They prey on aphids and leaf hoppers.

Chrysopa sp.: Its larvae feed on aphids.

Mantis: Feed on different insects.

Syrphids: They also feed on many insects.

Parasitic insects: The parasitic insects pass their life on the other insect and kill them. The important parasite of some insects are given below:

Trichogramma minutum — On sugarcane borers.

Bracon gelechiae — On pyrilla

Apanteles flavipes — On lepidopterous larvae.

Other Enemies: Besides, insect predators and parasites there are still other which prey on phytophagous insects, *e.g.*, frog, toad, reptiles, birds etc. Some mammals like rats, bears also prey on insects.

Cannibalism: The larvae of some insects feed on themselves and thus reducing the population, *e.g.*, *Heliothis* larvae.

Natural Diseases: Like other animals, the insects are also suffer from the diseases in great number and their population is checked is nature. The important micro-organisms which cause diseases are as follows:

(i) Protozoa

(ii) Bacteria

(iii) Fungi

(iv) Virus

(vi) Nematodes

The details of these micro-organism will be dealt in the chapter of biological control.

Artificial or Applied Control: Applied control includes those methods, under the control of man, which are necessary and used when harmful insects have not been held in check by natural agencies. The following factors are pre-requisites for the success of insect control:

(i) Exact taxonomy and morphology of the insect.

(ii) Life cycle, natural and seasonal history of the insect.

After having the understanding on above points we may know the damaging stages of insect, its habitat, nature and weak point in its life cycle, which will be helpful in devising the successful control strategy. Under applied control the following methods have been included:

(i) Mechanical and physical control
(ii) Cultural control
(iii) Legal control
(iv) Biological control
(v) Chemical control

Various Modes of Control

Mechanical and Physical Control

The reduction or suppression of insect population by means of specially designed machines or devices or by manipulation of physical factors of the environment are included under this group of control.

Mechanical Methods

The mechanical control incorporates the following devices:

Hand Collection of Insects: Hand picking of large sized immature or mature stages of insects is the most ancient method and this may prove very effective in the collection of egg masses and larvae of such insects which lay eggs in clusters and feed gregariously, *e.g.*, sugarcane top borer, Pyrilla, hairy caterpillars, grubs of mustard saw fly and various stages of Epilachna beetle.

The regular hand picking of rolled leaves, harrowing the caterpillars of cotton leaf roller or the dropping withered top shoot buds and bolls of cotton attacked by spotted boll worm are some of the example.

Hand Nets: Hand nets are usefully employed against the flying insects. They may be caught very easily during their migration when they are in large number, *e.g.*, grass hoppers, jassids, bugs, pyrilla and moths etc.

Mechanical Barriers: Barriers are used for preventing movement of insect which migrate by moving from one field to another the following are the common practices which generally followed:

(a) *Tree Banding:* Adhesive or sticky bands around the tree trunks are used against those insects which infest trees by climbing the trunks, e.g., Mango mealy bug.

(b) *Fencing:* It is possible to prevent invasion of a crop by migrating insects through the use of walls around the field so that may not go to another field.

(c) *Trenching:* The crawling herds of army worms may be stopped by constructing deep furrows, around fields against their marching ways. Trench barriers are also used to stop the locust hoppers.

(d) *Other Barriers:* Collar of individual plant to protect from cutworms; paper or muslin bags over fruits to protect from fruit flies, fruit sucking moths and anar butterflies etc., complete wrapping of tree trunks with paper against the flat headed borers are other examples of the use of mechanical barriers.

Mechanical Traps: Many injurious traps have been made to catch insects but a very large percentage of them fail to offer adequate crop protection.

(a) *Light Traps:* Light traps are generally used against those insects which are phototrophic, *i.e.,* attracted towards light. For this purpose a strong light may be placed or hanged in the field above a container having kerosinised water. A large number of insects attracted to light will be killed in the kerosinised water. Now-a-days, specially designed light traps to collect different types of insects are available and may be procured according to type and nature of insect. Light traps are not used only to attract and destroy the insect but also as indicator of the activity of pests in the field so that timely control operations can be undertaken.

(b) *Adhesive Traps:* Many adhesive traps are used to kill insects and for the purpose the pieces of bags or cloths are dipped in gum solution or in any other adhesive material having insecticide. These pieces are hanged on the trees in the vicinity of fields, the flying insects will come across and attached with the trap and finally die. Some times an attractant is also used in these adhesive traps to attract the insects.

(c) *Hopper Dozers or other Dozers*: Such type of traps are made for catching insects of a particular group and these may be handled by man, animal or by a tractor. Some commonly used hoppers are hopper catcher, aphidozers, fly traps, moth traps, maggot traps and electric traps etc.

Physical Methods

Physical method involves specially manipulation of change in temperature or humidity or employ radiant energy in some way to destroy pests:

Use of Temperature: Artificial heating or cooling of stored products, the mills or factories where such products are processed, is a common method of insect control.

(a) *Low Temperature:* Nearly all insects become inactive at 60-40°F. Few insects are not killed at these temperatures unless exposed to them for a considerable length of time. Generally insect undergo hibernation at 20 to 30°F. It is not certain that exposure to such temperature will kill the eggs of such species as the grain weevils. Practically no damage occur from insects at temperature below 40°F. Although low temperature is not as effective as high temperature in killing insects but storage of food products or clothing, the temperature at below or near freezing point will prevent all insects damage.

(b) *High Temperature:* It has experimentally been shown that no insect can survive longer when exposed to temperature of 140-150°F. Most of the insects including those which attack stored grains can be killed within 3 hours exposure to temperature from 125-130°F. In many countries mills and large cultivators have equipped their buildings with enough heating pipes to enable them to raise the temperature 125-150°F for several hours and thus kill all insects in the buildings. In tropical countries like India where the sun is strongly radiant during the greater part of year, solar radiation can be most economically employed in the case of insect infested materials such as cotton seeds, grains, pulses, spices, nuts and woollen goods etc. Super heating with burning charcoal can also be practiced to raise temperature in godown to destroy most of the stored pests.

Humidity: Some crops can be protected from insect pests either by flooding or by irrigation, at proper time. The draining of marshy land and other standing water is the most effective method of destroying mosquitoes and house flies. Drying of grains (less than 10% moisture) before storage is a sound practice to prevent insect damage by stored gram pests.

Light: The use of radiant energy to control insects has been experimentally studied by a host of workers. The light has been utilized to attract many strongly phototrophic insects into traps from which they can not escape. In general insects appear to be strongly attracted at about 3650 A, hence ultraviolet lamps have been used in some countries to trap the flying moths. The phototrophic response declines sharply in the violet region at 4400 A, minor attraction may occur at 4920 A (blue to blue green) 5150 A (green) or 5550 A (yellow green) but the longer wave length of 5560 to 7000 A (yellow to red) are relatively unattractive.

Ionizing radiation or radio activity has also been used for insect control the radio activity consists of particles, B particles, neutronts, X-rays and r-rays. Of these r-rays are extensively used because at much lower dosages these rays are capable of sterilizing insects. This method of controlling insects (sterilization with r-rays) has been used by knipling Lindquest and Bushland for the control of screw-worm fly in U.S.A. in 1937. Cobalt-60 source of r-radiation was used by these workers to sterilize the pupae which developed into adults and mated normally. The males require comparatively lower dosage (2500 r-) than females (5000 *r-)* for sterilization. This technique was successfully applied in 1958 to an area of about 64000 sq. miles of Florida where after the release of nearly 3 billion sterile flies over a 17 months period, eradication of screw worm flies was achieved. This technique (sterile male technique), no doubt, is applicable for the control of other insects also but the following principles have been established for its successful augmentation:

(i) Mass rearing of the insect must be possible.

(ii) Adequate dispersal of released sterile males must take place.

(iii) Mating behaviour of males should not be impaired by sterilization, *i.e.,* mating should be normal.

(iv) Female should normally mate only ones or if more matings occur, the sperms of sterile males must compete with those of normal.

(v) Initial population of insect should be checked so that there is no competitions between sterile and non-sterile males.

(vi) The sterile males released in fields should not damage crops.

Sound: Little work has been done on the use of sound waves in insect control. Adams and co-workers (1953-54) have employed these waves produced by feeding grainery weevil larvae to find infestation Frings and co-workers (1948) have killed insects with very high intensity of sound. The male mosquitoes are attracted by special type of sound and made sterile by electronic flash gun in Canada.

Cultural Control

Cultural control may be defined as control of a pest by slight variation, introduction or suppression of farm practices which are normally adopted in the cultivation of a crop. These practices may lead to the control of insect pest either by directly affecting their growth and multiplication or by minimizing the chance of their attack on plant through careful manipulation or through increase' in the internal resistance of plant. They must usually be employed well in advance of time when damage by the pest starts. This is the cheapest and practical method of control as it does not employ an extra money to be spent.

In order to control insects by cultural practices, it is necessary that one should know the life history and habits of insects so that the weak point of their life cycle may be pin pointed. A cultural practice effective against one kind of insect species may not be useful against a closely related species because of the difference in their habits. The cultural operations should be followed at proper stage of development of an insect. The following are the important cultural practices employed for insect control:

Ploughing or Cultivation of Soil: Insects are largely affected by the texture of soil, its chemical composition, percentage of soil moisture, the temperature and other soil organisms. Therefore, the various methods of soil management have a definite effect upon many insects. Deep ploughing of fields is a good method of controlling many insects particularly those which pass some portion of their life cycle in the soil. This method is recommended for hairy caterpillars, cut worms, white grubs, sphinged caterpillars etc. Periodically, breaking up of field bunds is useful to destroy grasshopper's eggs. Similarly racking and hoeing of soil around cucurbit plants and fruit trees is effective for destroying under ground pupae of fruit flies and also the red pumpkin beetle affecting vines.

Sanitation or Destruction of Crop Residues and Weeds: Crop residues and other traces remaining in fields after harvest provide food or shelter to many insect pests. The destruction of all stables after harvest of crops like paddy, maize, jowar, sugarcane, brinjal, cotton, sweet potato, and cucurbits etc. would prevent the multiplication of borers and other pests which would otherwise go unnoticed and appear on crop in the following season. The weeds growing around the cultivated field also serve as the host of many insects like the spotted boll worm and cotton leaf roller develop on many weeds growing in and around orchards. Hence the removal and destruction of these weeds is the only known practical method to prevent multiplication of these pests. Many insects hide or hibernate under loose bark in deep or unhealthy branches of trees. Removal of such shelter by scraping, pruning or plastering would prevent the multiplication of several pests such as bark eating caterpillar and shoot-stem borer.

Variation in the Time of Planting and Harvesting: The time of planting has a marked influence on the infestation of crops by some insects. By changing or carefully selecting the time when a crop is planted, we may avoid the egg laying period of a pest or get young plants well established before the infestation starts, *e.g.*, cucurbit vines sown early are better able to withstand the attack of red pumpkin beetle than those that are sown late and are still in the seedlings stage at the time of emergence of the beetle towards the end of February. Similarly, early sown cotton is found to be less liable to the attack of jassids than late sown cotton. This is also true with crops like mustard against aphid, linseed against gall midge and sesamum against jassids.

Crop Rotation: Crop rotation is one of the most frequently recommended methods of insect control. If a similar crop is grown year after year on a large area the insects of that crop predominate as they are getting continuous food. But if the second crop is grown next year, the insect dependant on the first crop may not be able to live on the second. Therefore, the crop planted in the field which was not grown previous year, will at least temporarily escape damage from its specific pests.

Among insects that injured cultivated crops the number of general feeders (polyphagous) is very small, those which feed on the plants of one family are numerous, and there are many that feed on only a few very closely related species. Therefore, much can be accomplished in preventing most insects from becoming serious pests in our fields, if a good rotation is practiced where a crop of one plant family follows to that of a different family. Crop rotations are generally more effective for insects those are restricted feeders, have limited powers of migration of sluggish habit and are slow breeder spending a relatively long time in the feeding stage.

The use of Resistant Varieties: Resistance of plant to insect attack may be defined as the relative amount of heritable qualities possessed by the plant which influence the ultimate degree of damage done by the insects. The intensity of resistant varies widely from crop to crop with associated insects. Some varieties or strains of cultivated plants are more or less resistant to some of the insects that attack them, hence the breeding of such resistant varieties of important crop plant has become a major weapon against insect attack. In last decade attempts have been made in abroad as well as in India to breed resistant varieties against variety of insect pests. But very little success has been achieved in this direction due to change in their feeding habit, existence of bio-types of particular insect and lack of complete resistance in the plant type. With the formation of various All India Coordinated Crop Improvement Projects suitable conditions have been created for under taking this type of work.

To develop resistant varieties studies were made on 100 insects and their host plants in U.S.A. and as a result of this only six varieties against hessian fly and European corn borer could be developed. During the last 30 years attempts were made in Punjab to evolve resistant variety of cotton against jassids but no success could be achieved. Even U4 variety of South-Africa which is resistant against the said pest failed to survive here. In our country concerted attempts are being made to develop the resistant varieties of sorghum, maize, paddy, cotton, lady's finger, cucurbits, mustard, sugarcane etc. More than 10,000 varieties from the world germ plasm collection have been screened against sorghum shoot fly and 10,196 varieties were tested against stem borer out of which 26 lines were indentified showing relatively less damage by this insect. About 2000 varieties of cotton belonging to 4 groups namely, arboreum, herbaceum, hirsutum and barbadense have been screened against jassids, boll worms and stem weevil at many locations in India. Generally, these varieties introduced from USA are susceptible to jassid. However, some cambidian types like B. 1007 and Bandawar and crosses between these varieties are reported to be resistant to jassid. Out of the 364 lines of pea tested against pea-pod borer only 3 lines were noted promising. The classical examples of some resistant varieties against particular pest are given in the following table:

S.No.	*Crop*	*Varieties*	*Insect*
1.	Apple	Winter majetin	Wooly aphids
2.	Grape	American Varieties	Phylloxera
3.	Wheat	Pawnee, Ponca, Ottawa	Hessian-fly
4.	Jowar	Atlas, sargo	Chinch bug
5.	Alfalfa	Ladack	Pea-aphid
6.	Alfalfa	Cody Lohantan	Spotted alfalfa aphid
7.	Maize	Ks-2234	Grasshopper
8.	Maize	Ks-1859	Corn leaf aphid
9.	Barely	Dictoo	Green bug

Other Cultural Practices: The following practices are also used in this method of control to regulate the activities of some insect pests:

Mixed Cropping: Due to sowing of mixed crop, the distance among the same kind of plants is increased, therefore, insects do not reach easily to them and thus resulting poor infestation. If jowar and bajra are sown mixed the attack of jowar stem borer is not so severe. Similarly, the sowing of wheat, gram, and mustard minimises the attack of termite, gram caterpillar and aphids on their respective hosts.

Some insect pests require two kinds of host to complete their seasonal cycle. This is particularly true with certain aphids in kumaun hills. Under such conditions the alternative host should be destroyed to check the multiplication of pest.

Only healthy and non-infested seeds or cuttings should be shown to check the carry over of the insects. This practice is particularly important in the case of sugarcane which may contain borer larvae and mealy bugs, cotton seeds which may have the pink boll worm larvae and cuttings of various garden plants which often contain mealy bugs and aphids etc.

Proper prunning and thinning, conservation and regulation of soil moisture, good preparation of seed bed and the judicious use of fertilizers offer possibilities of stimulating plant growth in such a way so as to grow profitable crop with minimum insect damage.

Removal of dry cane leaves to destroy the egg masses of pyrilla during winters and earthing of cane plants to prevent the attack of borer are some of the other useful cultural practices.

Sowing of trap crop is also another method to destroy and check the infestation of insect pests. This is particularly true with cotton when okra is grown to reduce the infestation of red cotton bug.

Legal Control

In the early days of agriculture development, the plants and their products were imported or exported with little or no thought to the insects or other pests that might be transported along with them. It is only in the middle of the 19th century when some countries thought about the legislation to restrict or prevent the spread of insects and other pests. The introduction of grape phylloxera from America to the grape-wine yards of France in 1860 caused such a serious destruction to the France vine that it was felt imperative to do something to prevent the unrestricted movement of infested grape vine to other part of the world. In

1881, the representatives of many European countries where grapes were extensively grown met and agreed to impose regulation restricting the movement of infested grape vines. Similarly the spread of sanjose scale in U.S.A. in later part of the 19th century laid to the passage of insect legislation in that country in 1905. In India, plant quarantine rules and regulations were made under the provision of the Destructive Pests, Plant Diseases and Noxious. Weeds Act No. 11 of 1914 in the guidance of Directorate of Plant Protection, Ministry of Food and Agriculture, Storage and Quarantine, Government of India. Before the enforcement of legal control following three important insects entered in the country:

Icerya Purchasi: It is a pest of citrus and came from Australia.

Aspidiotus Perniciosus: It attacks apple and pears and entered from China.

Eriosoma Lanigerum: It is also a pest of apple and pears and came from European countries.

The insect legislation may be broadly grouped into 5 classes:

(i) Legislation to prevent the introduction of new pests from foreign countries.

(ii) Legislation to prevent the spread of established pests within the country or within the state.

(iii) Legislation to enforce the application of control measures which have been found effective in preventing damage by established pests.

(iv) Legislation to prevent the adultration and misbranding of insecticides and to determine the impermissible residue tolerance limits in food stuff.

(v) Legislation to regulate the activities of pest control operations and the application of hazardous insecticides.

Quarantine Laws: The quarantine laws are the regulatory measures dealing with the introduction and spread of injurious insects and other pests. In India this responsibility rests with the Central Government under the provision of Destructive Pest and Insect Act of 1914, with the following rules:

(i) Prohibiting or restricting the import of plants and plant materials, insects and fungi into India.

(ii) Prohibiting or restricting the movement of insects or diseases and their hosts from one state to another state in India.

Under these rules, plant and materials are to be imported through one of the following prescribed ports of country:

Sea Ports: Mumbai, Kolkatta, Cochine, Dhanukushi, Kenkush and Bhava Nagar (only for American and West Indies cotton), Naga Patnam, Vishakhapatnam, Trivendram (only for unmanufactured tobacco), Chennai and Tuticorin. The ports of Chennai, Mumbai and Cochine have fumigating chamber's where fumigation is done.

Air Ports: Amritsar (only for fruits and vegetables) Mumbai, Kolkatta, Chennai, Palam and Safdarjang Delhi and Trichurapalli.

Land Frontiers: Khalra and Husainiwala for Pakistan and Guwahati for Burma and Bangla Desh.

All plants and plant materials are subjected to inspection and treatment, if necessary, at the port of entry. An official phytosanitary certificate in the International from the country of origin should accompany the consignment.

The plant quarantine department inspects the materials at three places:

(i) At the place from where material is being sent.

(ii) At the place where it reaches.

(iii) At the place where it is being kept.

The quarantine laws are so strictly followed that even beneficial insects may not be imported except at the following institutes:

(i) Indian Agriculture Research Institute, New Delhi.

(ii) Forest Research Institute, Dehradun.

(iii) Indian Veterinary Research Institute Mukteshwar.

(iv) Zoological Survey of India, Kolkata.

(v) Commonwealth Institute of Biological Control, Bangalore.

Complete restriction has been imposed on the import of following items as there is a chance to have such insects and diseases which are not found in India:

(i) Unginned Cotton, from Australia and Philippine, Newgni and Fijo islands.

(ii) Potato, onion, garlic and unmanufactured tobacco.

(iii) Unfumigated American Cotton.

(iv) Sugarcane, Seeds of Berseem and Sunflower etc.

Domestic quarantines exist against *Icerya purchasi,* bunchy top of banana, banana mosaic, sanjose scale and potato wart.

Insecticide Laws: Before the enforcement of Insecticides Act, 1968, pesticides were included under the Indian Poison Act, 1919, and Drug Act, 1940. Thereafter the Food Adultration Act, 1955 and Pesticide and Appliances Act, 1964 were enforced to regulate the standards of pesticides and appliances through Indian Standard Institute, New Delhi and Pest Control Committee. The standard materials are blended with ISI mark and these organisation also verify the information given on the labels pasted over the containers, equipments and appliances. In order to regulate the import, manufacture, sale, transportation, distribution and use of insecticides so as to prevent the risk to human beings or animals the following rules were framed in 1972 under Insecticide Act, 1968:

1. All the insecticides which are to be exported or manufactured must be registered at the central level. For this the registration form is to be submitted to the Insecticide Registration Committee at Directorate of Plant Protection and Quarantine, Faridabad. This

application form along with the registration fee of Rs. 100/- through a chalan is submitted under the head 065. Other administration services, other receipt fees, fines etc.

2. Any State can have the licence for formulating or selling the insecticide only after registration.
3. The State Government appoint Licensing Officers who renew or issue the licence for the formulation and sale of the insecticides.
4. For the renew or grant of a new licence the following fee is deposited in the treasury through chalan under the head 105-Agriculture (Gha) other receipt (4) other items
   (a) For the formulation of the insecticide the minimum fees is rupees 50/- and maximum is Rs. 500/-.
   (b) For the sale, stock, demonstration and distribution of each insecticide the minimum and maximum fee are Rs. 20/- and 300/- respectively.
5. The Central or the State Government appoint insecticide inspectors in specified areas for the following duties:
   (a) To inspect the sale points of his area jurisdiction at least thrice in a year.
   (b) To make sure whether the provisions of this act or rules made thereunder are being complied or not.
   (c) To collect the samples of such insecticides which he feels substandard or do not meet out the requirments of the Act or rules made thereunder and are being sold in the market.
   (d) To verify the written complaints and to resolve them.
   (e) To exercise such other powers as may be necessary for carrying out the purpose of this Act or the rules made thereunder.
   (f) To maintain the records of the samples brought from the different sale points and to keep the

inspection reports and to submit these reports to the officer concerned.

(g) To inspect and enquire about the sale and use of insecticides which are against the rules of insecticide act.

6. For the inspection of insecticide formulations following specific duties have been assigned to the Inspectors:

(a) To check the licences issued for insecticidal formulations in his area jurisdiction atleast thrice in a year and verify whether the terms and conditions mentioned in insecticide act are being followed or not.

(b) To submit the report just after the inspection regarding the follow-up of rules and regulations concerning the insecticides and their formulations.

(c) To take the samples of any insecticide and send such samples to the Insecticide Analyst for test in the prescribed manner.

(d) To report the Government about the cases of insecticidal poisoning.

(e) Period of the validity of the licence:

(i) The licence will be valid for two calendar years from the date of issue or its renewal,

(ii) Application for the renewal of licence should be given before its expiry and the following amount will be charged after the date of expiry.

(a) For the licence of insecticidal formulation Rs. 50/-.

(b) For other licence Rs. 20/- only.

(f) *Appeal:* Any person aggrieved by a decision of licensing officer may appeal to such authority (Director Agriculture) in such manner or payment of Rs. 10/- by treasury chalan within a period of 30 days from the date on which the decision is communicated to him.

7. There is a provision for Insecticide Analyst and Laboratories at State and Central level under this Act.
8. Packing and labelling of insecticides as per rules has been prescribed in the Act.
9. Provision has also been made for precautionary facilities, *viz.*, clothes, gas masks, gloves etc. to the persons handling insecticides during transportation and storage.

Compulsory Clean-up Measures: All the state Governments have legislation for the control or eradication of insects, other pests, diseases and weeds. Provision under the act authorise the restriction or prohibition of insect, destruction of insects and other pests, diseases and weeds and the plants or plant materials which harbour them. Areas or crops can be declared under the provision of laws as "pest infested or disease infected". Rules are made operative from time to time to deal with serious situation of insect pests, diseases, and weeds. The operation of the 'State Insects Diseases and Noxious weeds Act is the responsibility of the government through Agriculture and Revenue departments. An Act namely "Cotton Pest Act" was enforced in 1919 in Madras under which it was made compulsory that all farmers must destroy the residues of old crop by the end of August I of each year. Similarly, with the enforcement of Baroda State Agriculture Pest and Disease Act, 1940, it was necessary to destroy the stubbles of cotton by the end of 25th May. The following were the rules made by different states in respect of compulsory clean up measures:

1. The coorg Agriculture Pest and Disease Act, 1973.
2. The Patiyala Destructive Insect and Pest Act, 1943.
3. The Bombay Agriculture Pest and Disease Act, 1947.
4. The Rewa State Agriculture Pest and Disease Act, 1944.
5. The East Punjab Agriculture Pest, Disease and Noxious Weed, Act, 1949.
6. The Assam Agriculture Pest and Disease Act, 1950.
7. The Uttar Pradesh Locust Destruction Act, 1951.
8. The Bihar Pyrilla Destruction Act, 1953.

The person living in these states are bound to follow the instructions mentioned in the Act and failing to do so they are punished as per rules mentioned therein.

Biological Control

The use of natural enemies for the control of harmful insects, other animals and plants is known as biological control. The term biological control was first used by Smith 1919 to signify the use of natural enemies. In other word collecting and rearing of natural enemies in great numbers in the laboratories and releasing them on harmful insects, other animals and plants is known as biological control. In general we can say that the study and utilization of predators, parasites and pathogens for the control of injurious insects is known as biological control. Following are the natural enemies which are used in this way:

1. Predatory and parasitic insects.
2. Predatory vertebrates birds. reptiles, mammals etc. — Macrobial control
3. Parasitic nematodes
4. Protozoal diseases
5. Parasitic fungi — Microbial control
6. Bacterial diseases
7. Viral diseases
8. Rickettsae

Following are the four principles employed in the biological control:

1. Collecting parasites, predators and pathogens from the places where they naturally assembled in great numbers and releasing them in places where they may do most good.
2. Collecting and handling the host insects in such a way so as to kill them but permit any parasites or predators among them to escape.

3. Rearing of predators and parasites in the laboratories in great numbers and releasing them in such places where the population of host is reached to its maximum.
4. Importation of natural species from other places (foreign countries) and use them against the pests established from the foreign country.

Basic Steps of Biological Control

Following are the steps which have to be followed for the successful biological control:

1. Exact determination of controlling insect species or race to be controlled.
2. Careful study of the geographical distribution of the pest species.
3. Knowledge about the biology of the host insects and their natural enemies, parasites attacking on different developmental stages of the host insect.
4. Thorough knowledge of the indigenous parasites and predators of the insect pests.

Careful study of the natural enemies that they are entomophagous and also ensure that they are predatory, parasitic, if not, they may not be used.

Attributes of Effective Natural Enemies

A good natural enemy may have the following qualities:

1. The most effective attribute of a natural enemy is its searching capacity of their host. When the host population is scarce it should be able to find the host.
2. It should be highly specific or monophagy but do not pass their life on other hosts.
3. High fecundity, short life cycle, so that they may be produced in large number in a short period. More generations of the enemies may be developed to the generation of the host.

4. Adoptability to new environmental conditions.
5. Resistant to pesticides.
6. Free from hyperparasites.
7. Power for dispersal at least as much as the host.

History of Biological Control

Since ancient times; Horticulturist of Asian countries is using predaceous ants for reducing the number of foliage feeding insects of horticulture crops. But insect parasite *Apanteles glomeratus* was observed attacking on cabbage butterfly *Pieris rapae* L. which was first described by Vallisniesi in the year 1661. After that little attention was paid in this direction. In 1800 Erashus Darvin suggested in his book Phytologia that the predators and parasites may be used for biological control. After which scientists started thinking in this direction and made successful experiments. The remarkable success in this direction was the control of cottony cushion scale *(Icerya purchasi)* by the import of a lady bird beetle *(Rodolia cardinalis)* from Australia in California (USA) in 1888. This lady bird beetle was used in 32 countries of the world where it established well and completely checked the population of cottony cushion scale *(Icerya purchasi Mash).*

After that *Georgh Campier* 1890 made survey of different countries of the world and collected predators and parasites of fruit and scale insects and sent to America.

In 1906 a *Hawaiian Frediricik Mure* surveyed the natural enemies of sugarcane pests, and found Scolia *moniliae* was successfully controlling the *Anomala orientates.*

Agents of Biological Control

Vertebrate Predators: In this category different types of animals found to catch and devour the insects, *viz.,* Fishes, frog, toad, tortoise, snakes, lizard, birds (Hudhud, Mama, Blue-Jay, Partridge, Quail, Peacock, Papiha, Owl, Bhujanga, Cuckoo, Crow, Sparrow and Kite) mammals (Bats, Rats, Mole and Mongoos). Among the predatory vertebrates birds are most important but they are less used to introduce from one country to another. Indian maina was introduced from India to Hawai island. It is clear from these examples that these natural enemies are doing as a autocidal control of insect in the nature but they are not being used as biological control agents.

Predatory and Parasitic Insects: They are also controlling the insects in the nature. By seeing their effectiveness in nature some natural enemies are reared in the laboratories and utilized for the biological control of insects. In America from 1883 to 1894, 403 natural enemies were introduced from other countries in which only 94 established in the country. In India also a large number of predators and parasites have been introduced from different parts of the world of the biological control of harmful insects and only 12 are successfully utilized.

Predators are those insects which catch and devour the harmful insects while parasite makes their homes on or inside the body of the host and pass their life cycle and finally destroy the host insect. Some of the predators and parasites are given below which have been successfully utilized in different parts of the world of the biological control of insects.

Predator/Parasite	***Host Insect***
Rodolia cardinalis (Predator)	Cottony cushion scale *(Icerya purchasi)*
Coccophagus gumeyi (Parasite)	Cirtophilus mealybug
Tetracnemus pretisus (Parasite)	*(Pseudococcus gahani)*
Anarhopus sydneyensis (Parasite)	Long-tailed mealybug
Tetracnemus perigrinus (Parasite)	*(Pseudococcon-adonidum)*
Pseudopycus utilis (Parasite) mealybug	*(Pseudococcus-nipae)*
Cyrtorhinus mundulus (Predator)	Sugarcane leaf hopper (*Perkinsiella* saccharicida)
Microceromasia sphcnophori (Parasite)	Sugarcane weevil *(Rhabdoenemis obscura)*

Following facts are observed from study and utilization of natural enemies in the biological control:

1. 50% insects controlled are belonging to the Homoptera (Coccids, mealybugs, aphids, jassids and white flies) and rest to the orders Coleoptera and lepidoptera.
2. Success of the biological control is depend on the dominant natural enemy.
3. Four times more insects are controlled by parasites in comparison to predators.
4. Biological control is most successful in temperate region in comparison to tropical and subtropical regions.

Natural Enemies Imported in India

The following 13 natural enemies are found successful which have been imported from different parts of the world in India to control the insects given against their name.

	Imported predator/Parasite	*Host insect/Host-animal*
1.	*Rodolia cardinalis* *(Predator)*	Cottony cushion scale *(Icerya purchasi)*
2.	*Aphelinus mali (Parasite)*	Wolly aphid *(Eriosoma-lanigerum)* potato tuber moth
3.	*Bracon gelechae (Parasite)*	Potato tuber moth *(Gnorimoschema opercuella)*
4.	*Prospaltella Perniciosi* *(Parasite)*	Sanjose scale
5.	*Aphytis mytilaspidis (Parasite)*	*Quadraspidiotus-Pemiciosus*
6.	*Spoggosia beniziana* *(Parasite)*	Coconut caterpillar *(Nephantis serinopa)*
7.	*Opius vandenboschi (Parasite)*	Fruitfly *(Dacus* spp.)
8.	*Apanteles flavipes* *(Parasite)*	Jowar stem borer *(Chilo partellus)*
9.	*Telonomus nawai* *(Parasite)*	Castor semilooper *(Achoea janata)*

Contd...

	Imported predator/Parasite	Host insect/Host-animal
10.	*Trichogramma minutum* (*Parasite*)	Sugarcane and cotton borers
11.	*Lixophaga diatraeae* (*Parasite*)	Sugarcane borers
12.	*Euglandina rosea* (*Predatory snail*)	Giant african snail (*Achetina fulica*)
13.	*Syngamia haemorrhoidalis*	Lantana weed

Qualities of Introduced Natural Enemies

1. Ability to control such pest for which it was introduced.
2. It will never be become phytophagous.
3. It will not destroy the indigenous primary parasite and do not become hyperparasite.

Use of Micro Organism in Biological Control: Very small and microscopic component used in biological control are put into micro organism. That's why it is called as microbial control.

Bacteria: de Herelle 1911-1916 gave the 1st practical biological control of insects on the use of bacterium *Coccobacillus acridiorum* against locusts. From that time bacteria are being utilized for the control of harmful insects. In the beginning of 1940 in the United States, it was observed that milky disease A and B in Japanese beetle *(Popilliae japonica)* is caused by the bacteria *Bacillus popilliae and Bacillus lentimorbus.* These bacteria are successfully utilized for the control of Japanese beetle grubs. Infected larvae are crushed and mixed into the soil where they parasitise and kill the larvae. Now-a-days *Bacillus thuringiensis* is successfully used for the control of lepidopterous pests in different parts of the world. The various commercial formulations are available which are used to control the pests, *e.g.*, Dipel, thuricide, Biolip, Bio-tox etc.

Entomogenous bacteria may be classified as under:

Spore forming Obligate	Non spore forming Facultative	
	Potential	Facultative
1. *Bacillus popillae*	1. *Pseudomonas*	1. *Serratia marascens*
2. *Clostridium brevifaciens*	2. *Aerobactor*	

Crystalliferous	Non-crystalliferous
1. *Bacillus thuringiensis*	*Bacillus cereus*

Fungi: In the nineteenth century it was thought that the fungi may be utilized for the control of insects. From that time the fungi are being used for the control of injurious insects. These fungi attack the host body externally or internally where the spores germinate and cover the whole body and kill the host. In America, white muscardine fungus *Beauveria bassiana* has been successfully used for the 1st time against chinch bug *(Blessus lucupterous). Spear and Colley* 1912 used the fungus *Entomophthora aulicae* for the successful control of brown tail moth. Other species of entomophthora are also used against other insects. *Empusa spp.* has also been used against houseflies and grasshoppers. All the spores of above fungi are mixed in water and sprayed on the crop. These fungal spores attach to body of host where they germinate and produce mycelium and hyphae and kill the host. High humidity is needed for the successful germination of the spores. Other fungi which are effective for the biological control of insects are *Isaria, Spororichum and Aspergillus* etc.

Virus: Although the important role of virus diseases as natural control factor has long been recognised. But, it is only in the last few years that the viruses have been used successfully in control programme. First successful establishment of an exotic virus disease was accomplished by Balch (1946), who introduced the polyhedrosis virus of the European spruce-saw fly *(Diprion heraynial)* into Newfound land. The first successful attempt to introduce a virus disease not present previously in the North American continent was made by Bird (1953) who applied a polyhedrosis virus of the European pine saw fly *(Neodiprion sertifer)* obtained from Sweden.

Now-a-days polyhedrosis virus *(Borrelina campeolis)* has been successfully utilized in California for the commercial control of alfalfa caterpillar by spraying of virus suspension containing SO lac polyhedra/c.c. @ 5 gallon/acre (Thompson and Steinhaus). Steinhaus also used the polyhedrosis virus successfully against *Helio this armigera and Heliothis virescens.*

Entomogenous viruses may be grouped into two categories inclusion viruses producing inclusion bodies in the host cell and non-inclusion viruses which do not do so. The inclusion viruses are further classified into polyhedroses which produce polyhederal bodies and granulosis viruses which produce granular bodies. Polyhedrosis could inhabit the nucleus and are called nuclear polyhedrosis viruses or the cytoplasm which are known as cyto plasmic polyhedrosis viruses. The inclusion viruses are more effective as insect control agents than the non-inclusion viruses.

Protozoa: There are so many parasitic protozoa as reported by *Tanada* 1959 "that the parasite protozoan's may be used successfully as a biological control agent because of their abundance" Due to various hinderances in their mass production not much work has been done on this aspect. It may be one of the causes that they produce cronic disease infection inspite of acute disease as a result of which the host insects may survive for a longer period.

In America *Perezia pyraustae* has been successfully utilized for the control of European corn borer *(Ostrinia nubilalis).*

Nematode: Some nematode species are also known which parasitize and destroy the insects., Glacer and Wilaox 1918 described in detail for using nematodes as a biological agents.

In 1959 Tanada also reported the presence of nematodes as international parasite on various beetles, grasshoppers, moths and other insects and recommended that they may be used successfully for the control of these insects. Now-a-days DD-136 *(Neoaplec-tana carpocapsae)* is used against the control of various insect pests. The nematode suspension is sprayed on the crops for the control of lepidopterous, Coleopterous and other insect species. In India it has been found successful against a number of pests. The other nematode species are Neotylenchus, Rhabditis, Mermis, Hercamermis Protrellus, Pseudonymus Sternienonema etc.

Problems of Biological Control

1. Biological control agents have narrow host range and control only a particular pest.
2. The effect is very slow and required a long time for the control.
3. Highly trained personals are required because living agents are used.
4. More member of biological control agents are required to use so that much money and time required.
5. Some times introduced agents are also parasitized by parasites that's why no success achieved.
6. Some agents may have special environmental conditions for its development and multiplications, therefore, require waiting period.
7. Due to lack of suitable environmental conditions the introduced parasites or predators do not adopt the prevailing conditions. More time is required to study the adoptability of that agent.

Scope of Biological Control

1. Using agents are no toxic and do not leave any harmful residues.
2. Effect of biological control is permanent type. Once introduced may work constantly in the nature for several years.
3. They do not harm to beneficial organism as like chemical control.
4. Some agents can easily be mass produced economically and may be utilized.

5. It is free from resistant problem as like insecticidal resistance.
6. Some agents may be used in combination with insecticides. If insect save from the attack of insecticide may be controlled by the natural enemies.

Chemical Control

The chemical which either kill the insects or repel them are used in chemical control. Insecticides are included under this group of control and are those substances which are used for protecting plants from harmful insects. The chemicals that kill the pests (any living organism which is harmful) are known as pesticides which may be classified as under.

S. No.	*Pesticides*	*Against which used*
1.	Insecticides	Insects
2.	Nematicides	Nematodes
3.	Acaricides	Mites
4.	Rodenticides	Rodents
5.	Molluscicides	Snails and slugs
6.	Fungicides	Fungal diseases
7.	Weedicides	Weeds

Insecticides

The detailed classification of insecticides has been shown in the chart. On the basis of entry in the insect body they may be grouped as follows:

1. Stomach poison
2. Contact poison
3. Fumigants

Similarly on the basis of their origin they may be classified as:

1. Inorganic compounds
2. Organic compounds of plant origin
3. Synthetic organic compounds

Inorganic Insecticides

These compounds were mainly used before the introduction of DDT and other synthetic organic compounds. Their use now is greatly restricted, and that too only for the control of some specific pests. The following compounds have been included in this category:

1. Arsenic compounds
2. Fluorine compounds
3. Sulphur and its compounds
4. Phosphorus compounds
5. Antimani compounds
6. Boron compounds
7. Mercury compounds
8. Thallium compounds

Arsenic Compounds: The various compounds of the element arsenic are the most important and widely used stomach poisons for insects. Some of these arsenicals meet nearly all the requirements for an ideal stomach poisons, except that they leave residue on the treated plants which may be dangerous to man and animals. The commonly used arsenicals are as follows:

Lead Arsenate

It was first used as insecticide in U.S.A. in 1892 for controlling the gypsy moth. It is generally used in two forms:

1. Acid orthoarsenate
2. Basic orthoarsenate

The commercial lead arsenate is a mixture of these two in which acid form predominates. Lead arsenate is relatively weak poison as compared to other arsenicals. It is unsoluble in water, therefore, possesses high degree safety to foliage, one of the advantages in using lead arsenate is that it only affects insects feeding on treated material, while other useful insects like parasites and predators remain unaffected. Lead arsenate may be applied either as dust or sprary, but more generally used as a spray. As a dust, it is generally diluted with from 2 to 20 parts of some inert carriers such as talc, hydrated lime or gypsum. As a spray 1.5 to 2 kg of lead arsenate is dissolved in 500 lit of water which will be sufficient for about a hectare. However, different concentrations have been recommended to various pests, *e.g.*, 0.25% spray for the control of lemon butterfly, epilachna beetle and red pumpkin beetle, 0.6% spray for the control of castor semilooper. An equal quantity of hydrated lime is added to the insecticide to prevent phytotoxicity to tender foliage.

Calcium Arsenate

It was used as insecticide as early as in 1907 but the compound contains so much water soluble arsenic that it burnt the foliage badly. Later, it was used in 1920. The commercial calcium arsenate are a mixture of tri-ca-arsenate, acid ca-arsenate with a considerable excess of lime and calcium carbonate. The safe calcium arsenate have basic calcium arsenate as a principal constituent which hydrolyse much more slowly. Calcium arsenate may be applied in the same way as lead arsenate but it is mainly as a dust and rarely as spray or in bait material because of relatively poor adhesive properties. It is cheaper than lead arsenate but more toxic than the former to the insects. The old calcium arsenate is more toxic to plants also and only resistant plants such as cotton, potato and apple withstand it. The major demerits of its are as follows:

1. It is not more stable.
2. The danger of burning foliage.
3. The toxicity to higher animals.
4. Its incompatibility with nicotine sulphate, soap, sulphur, pyrethrum and chlorinated hydro-carbon.

Paris Green

This was the first important stomach poison and was developed in U.S.A. in about 1865 to control the colorado-potato beetle. It is a acito arsenite of copper and is green coarse powder on account of its high arsenic content it is a quickacting insect poison but due to its great solubility in water, it may cause a severe scorching of the foliage. This insecticide is no longer used in India due to its high toxicity. It has now only a limited use as a mosquito larvaecide and for the control of snails.

Sodium Arsenate

This is very highly soluble form of arsenical, therefore, is mainly used in poison baits. It is extremely toxic to plants, therefore, extensively used as a weed killer. In the preparation of baits for the control of locust and grasshoppers, it is used with wheat bran in 2% ratio.

Basic Copper Arsenate

This compound has been developed comparatively recently as a safe and stable arsenical for foliage application. It had advantage of being quite insoluble and resistant to hydrolysis by water and is only slightly affected by exposure to carbon-dioxide. It is about as toxic as lead arsenate to a variety of insects.

Fluorine Compounds: These are derivatives of hydro-fluoric acid (sodium fluorite), Fluo-silicic acid (sodium fluo silicate, barium fluosilicate) and fluo-aminic acid (sodium fluo aluminate). The following are the main characteristics of these compounds:

1. They kill the insect more quickly than the arsenicals.
2. They are cheaper.
3. Less toxic to mammals.
4. Comparatively safer to plants.
5. They act both as stomach as well as contact poison.
6. Some cases they have repellent action also.

Sodium Fluosilicate (Na Si F)

It is a white granular, odourless powder containing 60.6% fluorine. This compound is also soluble in water, therefore, should not be sprayed on plants. It is widely used in poison baits to control grasshoppers, locust cutworms and crickets. Since better and more effective insecticides are available hence this insecticide is not much used now.

Barium Fluosilicate (Ba Si F)

Barium fluosilicate contains 40.8% fluorine and is least soluble of the inorganic fluorine, therefore, can be used in the form of spray. It is used at the rate of 1.5 to 3 kg barium fluosilicate in 500 litres of water per acre to control the flea beetles and blister beetles. Dust formulation may be prepared by mixing 1 part of it to 3 parts of inert dust viz. chalk, talc, clay or tobacco dust. It is reported to be corrosive to parts of spray machinery. It is incompatible with nicotine, calcium arsenate, lime sulphur, bordeaux mixture and alkaline soaps.

Cryolite or Sodium Fluo Aluminate (Na 3 Al F 6)

It is found in a natural form as a mineral in green land. In the mineral form, it is crystalline and contains 90 to 98% sodium fluoraluminate. The synthetic cryolite is lighter and fluffer than natural cryolite and contains 86% Na 3 Al F6. It is less soluble in water but soluble in dilute acid and alkalies. It is used as a spray at the rate of 2.5 kg in 459 lit of water per hectare. For dust preparation it is diluted with inert diluents such as clays and talc in the ratio of 1: 2 and used many chewing insects.

Sulphur and its Compounds: Sulphur is used against insects, mites and fungal diseases. Sulphur acts on fungi both by contact and by producing gases which have fungicidal action. It has little residual action, but due to following reason it is not used frequently.

1. Due to its odour.
2. Corrosive action on metals.
3. Bad effect on the colour of cloths.
4. Germination power of seeds is reduced.

It is used in the form of dust, wettable powder and spray as lime sulphur.

Lime Sulphur

Lime Sulphur is prepared as follows:

1. Sulphur 2 kg
2. Lime 10 kg
3. Water 50 lit

The lime is dissolved in big container having 1/3 of water (*i.e.,* about 16-17 lit). The sulphur is mixed gradually. This mixture is boiled for about an hour and remaining water is added. When the colour becomes brown it is allowed to cool. The mixture should not be heated more otherwise it will be highly phyto-toxic. 40 lit of this mixture may be mixed with 1000 lit of water to spray the solution in a hectare.

Lime sulphur may also be used in the form of the dust which may be prepared by mixing 1 kg sulphur in 5 kg hydrated lime this is used @ 25-30 kg/hect.

Phosphorus Compounds: Among the phosphorus compounds zinc phosphide is commonly used against grasshoppers and cricket. For the control of rats it is used in the bait.

Antimony Compounds (Tartar amatic): It contains antimonile acetate and used as bait poison in combination with sugar as attractant. Tartar amatic is also used to prepare poison baits against fruit flies and thrips.

Tartar acid 5 kg

Mollasses 25 kg

Water 1200 lit

The above quantity is sufficient for a hectare.

Mercuric Compounds (Mercuric chloride): It is so called as mercurous chloride or corrosive supplement or calomel. This compound is not in much use at present but was previously used against root affecting insect such as larvae of flea beetle, cabbage maggots and onion maggots etc.

Thallium Compounds (Thallium sulphate): It is used with some sweet as bait carrier to form a valuable poison against ants.

Boron Compounds: Amongst the boron compounds, Borex or sodium tetra borate is used to kill the maggots of house fly and mosquitoes in ponds and pits of farm yard manure.

Organic Insecticides of Plant Origin: The insecticides of plant origin extracted from seed, flower, leaves, stem and roots are termed as botanical insecticides. Unlike synthetic organic insecticides they are safer to use, but since they are expensive and lack residual toxicity, their use has been limited in India.

Nicotine: Nicotine occurs in atleast 15 species of *Nicotiana* but the commercially important species are *Nicotiana tobacum* and *N. rustica.* Nicotine commonly occur in *N. tobacum* plant to the extent of 2.5% in the leaves and 5 to 14% in *N. rustica.*

Tobacco was one of the first material to used as an insecticide as early as 1763. The use of finely powder tobacco mix with water was recommended to destroy aphids without injury to the foliage. In 1828, an alkaloid namely nicotine was recognised which was synthesised by Picted and Retschy in 1904. The alkaloid nicotine when freshly distilled is a colourless nearly odourless liquid boiling point 247°C, specific gravity 1.009 at 20°C and vapour pressure 0.0425 mm Hg. at 25°C. It can be purified by vacuum distillation. When exposed to air, its darkness changes from red to brown to black becomes more viscous and develops specific odour. It is soluble in alcohol, ether or petroleum ether and is miscible with water due to of its basic nature. Nicotine usually forms salts with acids and dibasic salt with many metals and acids. The nicotine sulphate $(C_{10} H_{10} N_2)_2 . H_2SO_4$ is widely used as an insecticide as it is less toxic to warm blooded animals than nicotine and is less volatile, therefore, more stable. When alkaline water or soap solution is added to nicotine sulphate, the free nicotine is liberated and shows higher insecticidal activity than the sulphate.

The nicotine enters in the body through the spiracles and acts on the ganglia of the nervous system. The less amount of nicotine produces excitement on the insect while more amount produces paralysis due to which death takes place. It can be used by various ways as follows:

(i) *As a Dust:* The concentrated nicotine extract is mixed with talc, gypsum and sulphur to prepare the dust. The type of carrier has an important effect upon the dust.

(ii) *As a Liquid Extract:* A decoction of tobacco leaves with soap is prepared by boiling 1 kg of leaves, stalks etc. of tobacco in 10 lit of water. One kg of soft soap is added after the water has boiled. This is continuously stirred and after a thick dark coloured liquid is obtained, strained. This stock solution should be diluted 10 to 15 times with water before spraying.

A standardized 40% nicotine sulphate is also available to use as spray which remains as toxic residue for a short time after application. The nicotine sprays are diluted so as to contain 0.05 to 0.06% nicotine. It has been successfully used in the control of sucking insects, *viz.*, aphids, thrips, psyllids, leaf miners and jassids.

(iii) *As Fumigants:* It is also used as fumigant either by vapourising the concentrated extract with heat or by burning papers or powders soaked in the extract.

Nicotine is compatible with many of the common insecticides and fungicides, including sulphur, lime sulphur and copper preparations.

Pyrethrum: The use of Pyrethrum as an insecticide originated in trans Caucasus region of Asia about the year 1800. For many years its nature was kept secret by these countries and the product was sold as flea and louce powder. Before IInd world war it was largely prepared in Japan but at present Kenya is leading in its production.

Pyrethrum occurs only the plants belonging to the genus Chrysanthemum = (Pyrethrum) family composite. The two species namely C. cinerariaefolium and C. coccineum possess a sufficient amount of toxic content. The former species is, however, of commercial importance. The active ingredients are:

1. Pyrethrin I, ester of pyrethrolone with chrysanthemum monocaroxylic acid.
2. Cinerin I, ester of cinerolone with chrysanthemum monocarboxylic acid.
3. Pyrethrin II, ester of pyrethrolone with chrysanthemum dicarboxylic acid monomethyl ester.
4. Cinerin II, ester of cinerolone with chrysanthemum dicarboxylic acid monomethyl ester.

The pyrethrin content of Indian flowers is 0.5% to 1.41% and inspite of 79 tonnes of flower production in Kashmir valley, 200-300 tons of flowers from Kenya which have 1.43-1. 83% pyrethrins are imported every year to meet indigenous requirement of pyrethrins. The cinerin is more stable than the pyrethrins and pyrethrin I and cinerin I seem to be more toxic than pyrethrin II and cinerin II, respectively.

Pyrethrum concentrates are prepared by ground flowers with solvents such as petroleum-ether, glacial acitic acid, methyl-alcohol or acetone. The pyrethrins are highly unstable in the presence of light, moisture and air. Whole flowers decompose more slowly than ground flowers or dusts. Pyrethrum preparations are rapid-acting contact insecticides, known for a quick knock-down effect on the insects. There is very limited residual effect and to enhance its toxicity and residual action, certain chemicals such as piperonyl butoxide are mixed to it. Pyrethrins are decomposed in alkaline solutions, therefore, they are incompatible with the lime sulphur, lime, calcium arsenate and paris green etc. Following preparations are generally available for use:

(i) *Dust:* It is prepared by mixing pyrethrin extracts with appropriate clay or talc carriers and formulated as 0.25 to 0.5% dust which may be used @ 25-30 kg/hect.

(ii) *Emulsifiable Concentrate:* Emulsifiable concentrate containing 3 to 20% pyrethrins is formulated which may be used as spray by diluting in water as 0.002 to 0.004%.

(iii) *Aerosole Bomb:* For use in aerosole bomb a 20% pyrethrin concentrate is used in such a way to give final concentration of 0.04 to 0.25% pyrethrins. It is used to kill the flying insects in the air.

Rotenone: Plants containing rotenoids were used as fish poison by the natives in tropical and some semitropical countries. But it was only in 1902, when rotenone was extracted from the roots of *Derris elliptica* and *D. malaccensis* in America. Rotenone has been extracted from the 68 species of plants including 21 *species* of *Teprosia*, 12 of *Derris*, 12 of *Lenchocarpus*, 10 of *Millattia* and several of *Mundulea* but two species of *Derris* are the economic one for commercial production. The rotenone content in the roots may vary from 4 to 11% depending on the variety. It is effective as a contact poison like pyrethrum, and can, therefore, be used against aphids, thrips, flea beetles and caterpillars etc., on vegetable crops. When exposed to light and air, it decomposes changing from colourless to yellow to deep red and resulting in non-insecticidal product. Rotenone is readily oxidized in the presence of alkali, hence incompatible with alkaline dusts and other alkaline wetting and spreading agents. Pure crystalline rotenone is prepared by extracting powdered rotenone containing roots with a solvent such as eather or carbon tetra chloride. It is harmless to human beings and animals but is extremely poisonous to fishes. It is soluble in ethyl alcohol, benzene, acetone, xylene, carbon tetra chloride and insoluble in water. It is formulated as below:

(i) *Dust:* Dust formulations are usually prepared by mixing dried ground roots with 3-5 parts of suitable carriers such as talc, clay, gypsum and sulphur etc. A dust containing 0.5% to 1.0% rotenone is generally used to control the insect pests.

(ii) *Wettable Powder:* Wettable or dispersible powder may be made from finally ground derris roots by mixing with natural soap 5% wettable powder is generally available, 2 kg of which is mixed in 500 lit of water to control the insects.

(iii) *Extracts of Derris:* These are available as dried reagents containing 25 to 30% rotenone and the rest amount of refined petroleum oils. Since rotenone has limited solubility at petroleum oils, natural solvents are generally employed to increase the solubility. The spray is used at a concentration of 0.001 to 0.002%.

Ryania

The root and stem of a South America plant *Ryania speciosa* have been found to have 0.16 to 0.2% insecticidal components, the most important of which is the alkaloid ryanodine. It is effective both as stomach and contact poison and soluble in water and most of the common organic solvents except benzene and petroleum oils. The ryanodine is more stable to the action of air and light than pyrethrum or rotenone and has considerable residual action which appears to be less toxic to mammals than rotenone. It is used as 20 to 40% dust and 0.3 to 0.6% spray to control a variety of pest. In foreign countries, Ryania is being used against borers of corn and sugarcane with benefit, but in India since it is imported and twice as costly as pyrethrum, hence its use is not so popular. Ld 50 = 750 mg/kg.

Hellebore

It is extracted from the roots of *Veratrum album* and *V. viride* which are incorrectly called as white and green hellebore, respectively.

V. *album* is native of Europe and Northern Asia while V. *viride* is found in U.S.A. The hellebore acts as stomach poison and specially used as larvaecide for house flies in manure. It is not in much use because of high cost, lack of standardization of toxic content and because only fresh ground material is effective.

Sabadilla

It is extracted from the seed of sabadilla plant *(Schoenocaulon officinale)* of South and Central America. The people of these countries have been using this as louse powder since long, but it was not until 1938 that a thorough investigation of their insecticidal properties were begin. It is present in the seeds upto 2-4% from which about 13% cevadine, ($C_{32}H_{49}O_9N$) 10% veratradine ($C_{36}H_{51}O_{11}N$) and lesser amount of civadilline, sobitine and sabidine have been isolated. Among these cevadine and veratradine are more insecticidally active and act as both contact, and stomach poisons. It is sensitive to light, therefore, should be kept in closed container in dark places.

Sabadilla is generally used as 10-20% dust or spray available in the form of 50% wettable powder. Its rapid deterioration upon exposure to light makes its use safer specially on crop at the time of harvest.

Quassia

It is found in the bark and woods of two trees namely *Quassia amara* and *Picrasma excelsa* from which Quassia and Picrasmin alkaloids, having same imperical formula ($C_{22}H_{30}O_6$) were identified. Decoction of Quassia in water with soft soaps have been found effective for the control of aphids and saw flies.

Scabrin

It is extracted from the roots of two weeds namely, *Heliopsis scabra* and H. *longipis* It is just like pyrethrum in nature and mostly used to kill the house fly.

Neem Seeds

The seeds of neem *(Azadirachta indica)* possess following four ingredients which have insecticidal properties:

(i) Azadirachtin
(ii) Nimbidine
(iii) Meliantriol
(iv) Nimbine

Suspension of powdered seeds at 0.1% concentration act as repellent for desert locust but the powdered material in the ratio of 1: 100 serves as protectant of grains against storage pest.

Anonaine

Powdered dried seeds, leaves, bark and roots of custard apple *(Anona reticulata, A. sguamesa)* have been found to exhibit insecticidal properties. The alkaloid has been identified as anonaine which acts both as stomach and contact poisons against insect pest.

Acorin

The roots of the plant sweet flag *(Acorus calamus)* produces acorin glucocide which has insecticidal properties. The insecticidal property of acorin has not been exploited fully but still the farmers of Orissa mix the powdered root with paddy for protection against storage pest.

Karanjin

It is furo-coumarin compound isolated from the acetone soluble fricton of the seed of Karanj *(Pongamia glabra)* which has manifested insecticidal property against various stored grain pests namely *(Tribolium castaneum, Sitophilus oryzae, Callosobruchus chinensis* etc.) Recently Karanjin has also exhibit, juveno mimmelic property against Indian Flesh fly.

Ricinin

The leaves of *Ricinus communis* contain the alkaloid ricinin which has been found toxic to the larvae of coddling moth.

Oils and Emulsions

The term oil is generic expression for a great variety of chemical substances which are composed of carbon, hydrogen, At ordinary temperature, they are greatly fluid insoluble in water, readily soluble in ether and chloroform etc. They are readily inflammable, slightly lighter than water and have peculiar property of penetrating into the pore of dry substances. There are mainly three classes of oils:

Fixed Oils: These are the animals fat and vegetable oils such as fish oil, linseed, soyabean, castor and coconut oil. They are essentially glycocides which saponify or form soap with alkaline bases. Fish oil and soyabean have been mostly used for making insecticidal soap. The soap may be prepared with the following:

Fish or Mahua oil	2 kg
Rosin	3 kg
Castic soda	650 g
Water	3.5 lit

1. kg soap after dissolving in 25 litres of water may be sprayed on a tree to control the nymphs of *Drosicha mangiferae*. For the control of sugarcane white fly 5 kg of soap should be mixed in 1000 litres of water and sprayed in a hectare.

Aromatic Oils: These differ from the fixed oils as being not greasy or viscous and not being capable of saponification. They derived from the plants and have a very pungent odour, e.g., citronella, camphene, menthal, piperment, euginol. Euginol is used to attract the fruit flies where as citronella to repel the mosquitoes.

Petroleum Oils: These are derived from sedimentary rocks and are complex solution of hundreds of hydrocarbons from which are derived natural gas, kerosine, gaseline, lubricating oil, asphatar and other mixture. As per origin crude petroleum oils are usually classified as paraphenic base, nephthenic or asphalitic base and mixed base oils. Actually all crude petroleum oils are mixture of all three types of components. Petroleum oils have been proved good contact poisons but are dangerous to green foliage; therefore, rarely used on plants. Light oils such as kerosine, dutrex or aromax are often used as solvents or diluents of insecticides. Petrol is popular for the control of stem boress of mango, apple and orange. Since the petroleum oils are harmful to plants, therefore, their emulsion is used. In the form of emulsion, oils are commonly used in combination with soap which acts as emulsifier. Diesel oil emulsion is one of the commonest emulsion used in India against san jose scale, which may be prepared as under:

Diesel oil	10 kg
Soft soap	4 kg
Water	28 litres

Diesel is generally added to the boiling soap water solution. The emulation thus formed is diluted with 30 parts of water before being applied to the trees.

Synthetic Organic Insecticides

Perhaps the 1st use of synthetic chemical was in 1892 when the potassium salt of 4, 6-dinitro-ortho-cresol was marketed in Germany as insecticide. However, it was the development of DDT during the IInd world war which considerably proved that a synthetic compound would proved superior to a variety of inorganic and natural products for many insecticidal uses.

Since then a large number of chemicals have been produced as effective insecticides. Synthetic organic insecticides may be grouped as follows:

1. Dinitro-phenols
2. Organic thiocynates
3. Chlorinated hydrocarbons
4. Organic phosphates
5. Carbonates
6. Synthetic pyrethroides

Dinitro-phenols: The earlier use of the dinitro phenols as insecticide was in 1892 and they are the derivatives of 4, 6-dinitro 2-alkylphenols and their salts or esters. These are generally used against scale insects, aphids, ticks and mites and include following insecticides:

DNOC: It is yellow crystalline solid (mp. 85°C) soluble in water to 0.014% at 15°C and forms salt with organic and inorganic bases. 4, 6-dinitro-ortho-cresol has the following structural formula:

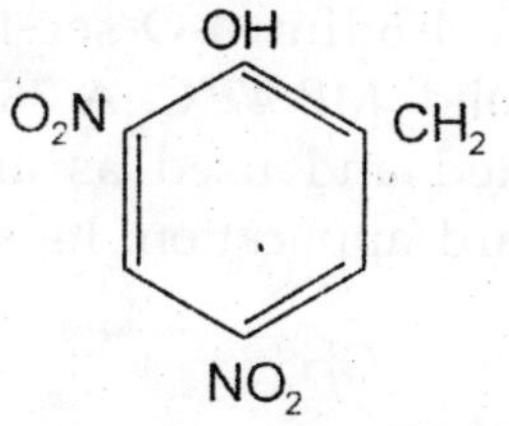

OH
O_2N CH_2
NO_2

The compound readily forms salt with Na, K, NH, C and Ba which are water soluble. DNOC is a powerful insecticide, ovicide and acaricide but its pytotoxicity has restricted its use to the control of grass-hoppers, locusts and over-wintering insects in the form of baits. In India 0.1% of its Na salt has proved effective in the control of sanjose scale in Himachal Pradesh. It is also used as weedicide. The acute oral LD 50 for rat is 40 to 65 mg/kg.

DNOCHP ($C_{18}H_{24}N_2O_6$): It is name given to 4, 6-dinitro-O-cyclohexylphenol. DNOCHP is yellowish white odourless crystalline solid, mp 106°C and only slightly soluble in water and solvents such as acetone, benzene, CCl_4, ethyl-acetate, ethylenedichloride, xylene, etc. Its structural formula is:

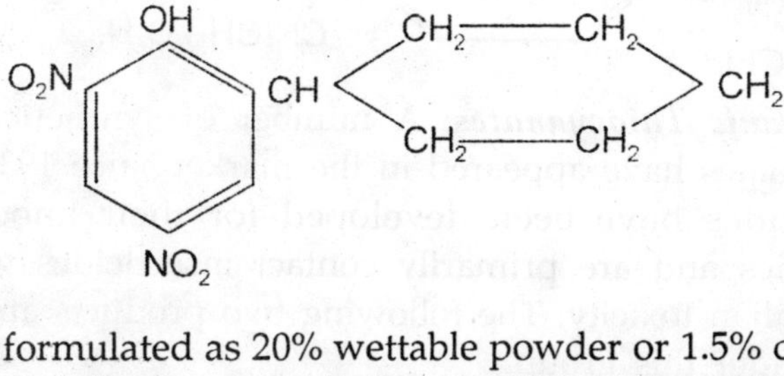

OH CH_2—CH_2
O_2N CH CH_2
CH_2—CH_2
NO_2

It is formulated as 20% wettable powder or 1.5% dust and mostly used for the control of red spider mites on fruit trees. It should be used with care upon all the plant foliage because of the injury which may result from over dose or unusual weather conditions of high temperature and humidity.

Both the above chemicals inhibit oxidative phosphorylation in insects and mammals and cause increased oxygen consumption which is responsible for toxic action of the compounds.

DNOSBP: It is 4-6dintro-O-sec-butylphenol and is crystalline orange solid, MP 42°C. A 36% solution of its salt in water is marketed and used as an effective dormant insecticide for orchard application. Its structural formula is:

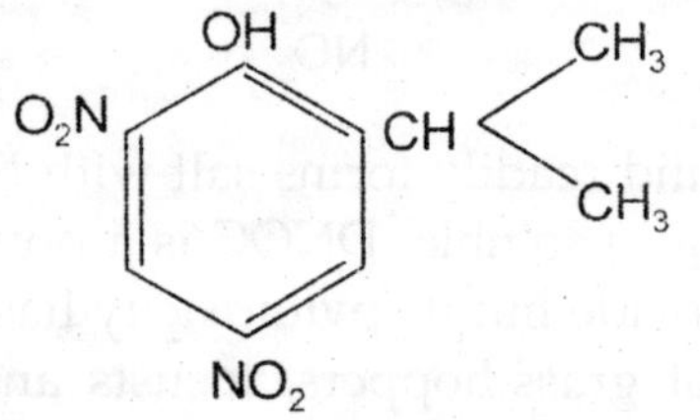

OH, O_2N, CH(CH_3)(CH_3), NO_2

DINOCAP: It is 2-(l-methyl-n-hepty)-4, 6-dinitrophenyl crotonate or 4, 6-dinitro-2-caprylphenyl crotonate. DINOCAP is commonly known as karathane or crotothane and used as an acaricide and fungicide available as 25% wettable powder. It has a low toxicity to mammals. Its oral acute LD 50 for rat is 980 to 1190 mg/kg. The structural formula is given below:

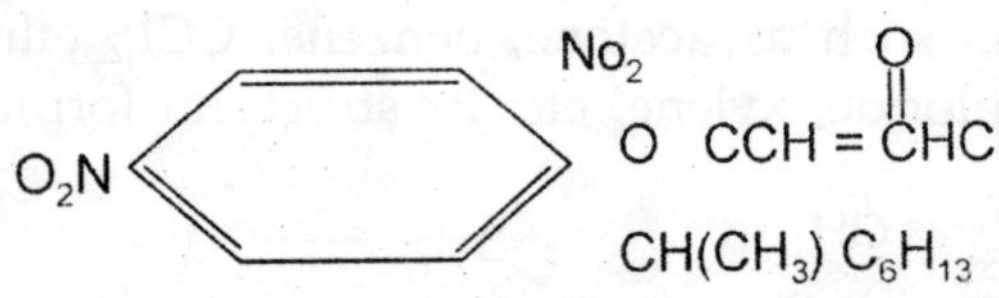

No_2, O_2N, $O\ \overset{O}{\overset{\|}{C}}CH = CHCH_3$, $CH(CH_3)\ C_6H_{13}$

Organic Thiocyanates: A number of synthetic organic thiocyanates have appeared in the market since 1932. These insecticides have been developed for their knock-down properties and are primarily contact insecticides with low mammalian toxicity. The following two products are mainly used under this group.

Lethanes: Louryl thiocyanate (C_{12} H_{25} SCN) is the main constituent of these thiocyanates. There are various trade names under which the lethanes have been formulated, *viz.*, lethane-

60, lethane 384, lethane A-70, lethane B-72 and loro. It is generally used for the control of aphids, aleyrodids, thrips, spider and mites etc.

Thanites: The thanites are produced by treating the camphene from palm oil or tarpint oil with chloroacetic acid and than with an alkayl thiocyanate. The technical product containing 80% isobornyl thiocyanoacetate and the remainder closely related bornyl and fenchyl thiocyanoacetates. It is marketed as a technical product as well as 20% solution in deodorised kerosine and mostly used as fly, cattle sprays and as household insecticide.

Chlorinated Hydrocarbon: The insecticides of this group have been developed after IInd world war and appeared in public with DDT. Organochlorine compounds have carbon, hydrogen and chlorine as their basic molecular constituents and some compounds may have oxygen and sulphur also. They may be grouped as under:

(i) Diphenyl methane structure	— DDT and its analogues
(ii) Monocyclic structure	— BHC
(iii) Bicyclic structure	— Toxaphene, Strobane
(iv) Polycyclic structure	— Aldrin, Chordane, Heptachlor
(v) Heterocyclic structure	— Endrin, Dieldrin, Thiodan, Telodrin

Diphenyl Methane Structure

These compounds have few benzene rings attached to a single carbon atom and include DDT and its analogues.

DDT

It is dichlorodiphenyl trichloroethane ($C_{14} H_9 Cl_5$) and was first synthesized by German chemist Zeidler in 1874, but its insecticidal properties were not discovered until the work of Paul Muller in 1939. It is produced by reacting chloral hydrate with monochlorobenzene in the presence of concentrated sulphuric acid. Technical DDT is a white to cream coloured amorphous powder and consist of a mixture of atleast 14 chemical compounds. These included, 65-80% of active pp-DDT, 15-21% of the nearly inactive op'- DDT, up to 4% of DDD or TDE and other related materials. The principal ingredient of technical DDT, pp'-DDT, is 1.11-trichloro-2, 2 di-(4-chlorophenyl) ethane or I, I, I trichloro 2, 2-bis (p-chlorophenyl) ethane. The structural formula is as follows:

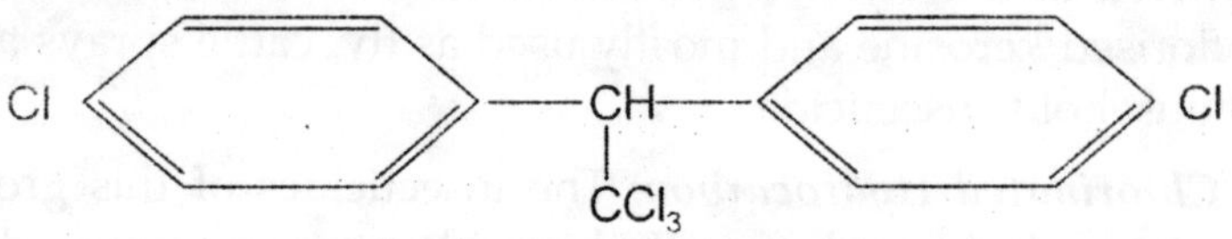

Cl —CH— Cl

CCl_3

The technical DDT has a setting point of 89% and should contain 9.5 to 11% hydrolysable chlorine and less than 1% volatile and alcohol insoluble material. DDT is almost completely insoluble in water (0.00001%), fairly soluble in most organic solvents and highly soluble in aromatic hydrocarbons such as xylene. In natural oils, e.g., linseed, pea-nut, cotton seed oil, the solubility is 10 to 12%.

It is readily dehydrochlorinated in alkaline solutions to form a non-insecticidal product namely dichloro ethylene. It is generally stable under most conditions and has a very low vapour pressure. DDT is a contact and stomach poison has long residual action. It affects the sensory organs and nervous system and causes violent agitation which is followed by paralysis and death. It is slow acting insecticide relatively nontoxic to mammals but in oil solution it is absorbed by the skin. When animals are fed with treated foliage, accumulation of the DDT in animal fat and its secretion in the fat (butter) of lactating animal has been observed. DDT is relatively safe for use on large number of crops but exhibits pytotoxicity on sensitive crops like cucurbits. It is also found to stimulate plant growth in some crops like potato, tomato, brinjal, cabbage etc. at lower concentrations. DDT is effective against a wide range of insect pests although some of them (singhara beetle, cutworm, sugarcane pyrilla and mango jassids etc.) have developed resistance to it. Since DDT kills natural enemies, hence some insects such as aphids, scale insects mealy bugs, grasshoppers and mites develop in more number after its application. The oral and dermal LD 50 for rat are 300-500 and 2500 mg/kg respectively. It can be used in combination with many insecticides except lime sulphur or any strong alkaline material in which it decomposes rapidly DDT is available in the following form:

Dusting Powder: It is white or dirty white in colour and mainly available in the strength of 2 to 10 per cent of the toxicant. The dust is prepared by mixing the chemical with inert material like talc, gypsum and pyrophillite, etc. It is applied generally at the rate of 25 to 30 kg per hectare against a variety of crop pests, *viz.*, swarming caterpillar, army worm, rice-case worm, cut worms, potato beetles, cotton leaf roller etc.

Water-dispersible Powder or Wettable Powder: It is used for preparing suspensions in water and is available in a white powder containing 50% of actual DDT. It is used as 0.15 to 0.32% spray for the control of most agricultural pests.

Emulsifiable Concentrate: DDT is available as 25% emulsifiable concentrate and dark brown in colour, It is used at 0.15 to 0.32% concentration for the control of pests. Due to larger residual action, emulsions preferred to the water-despersible powder.

Several DDT analogues which have insecticidal properties are also used to control insects. Among them TDE or DDD, perthane, methoxychlor, dilan, dicofal and DMC are most common.

Monocyclic Structure

These compounds contain only one non-aeromatic hydrocarbon ring structure.

BHC-BHC is 1, 2, 3, 4, 5, 6-Hexachlorocyclohexane which has the empirical formula $C_6H_6Cl_6$. Its common name benzene hexachloride is incorrect from a chemical stand point as HCH could be confused with hexachloro benzene. This compound was first prepared by Michael Faraday in 1825. Its insecticidal properties were discovered simultaneously in England and France in about 1942. Its structural formula is as follows:

Cl

Cl Cl

Cl Cl

Cl

It is produced by the chlorination of benzene in the presence of sunlight. The technical product is greyish or brownish amorphous solid and has a strong musty odour. Among the 13 possible isomers found in varying proportions in BHC, the gamma isomer is insecticidically most active and the percentage of following 6 isomers in BHC is as follows:

Isomers	*Melting point in °C*	*Percentage in BHC*
Alfa	157	55-77
Beta	309	5-14
Gamma	112	10-18
Delta	138	6-8
Epsilon	219	3-4
Eta 90-91	Unknown	

The 6 isomers differ widely in their solubilities in the common organic solvents, in their degree of volatility and in their stability to alkali. BHC is stable compound but is broken down by alkali. It acts as a stomach and contact poison and also to some extent as a fumigant. BHC is more toxic to insects than DDT but less poisonous to man and domesticated animals than DDT. BHC poisoning increases respiration rate with the symptoms of tremors, ataxia, convulsions and prostraction. BHC is more volatile than DDT and therefore, does not have a long residual toxicity. In closed places it is effective about a week and in field 3-4 weeks. It is available in the following formulations.

Dust: The dusting powder of BHC contains five to ten percent of the active ingredient and is commonly used @ 25 to 30 kg/hect to control a variety of crop pests. The dust is light grey or light brown in colour.

Wettable Powder: The water-dispersible powder is available in 50% concentration and light brown in colour. It is used in spray in the concentration of 0.15% to 0.5% against aphids, jassids, white flies, leaf rollers, leaf cutting beetles and thrips etc.

Emulsifiable Concentrate: Its emulsion is dark brown to light brown in colour and is available as 20% emulsifiable concentrate. It is used in the same concentration as wettable powder against the same pests.

BHC can be used with other insecticides and fungicides except those having strong alkaline materials, *e.g.*, Bordeaux mixture, lime or lime-sulphur.

Lindane

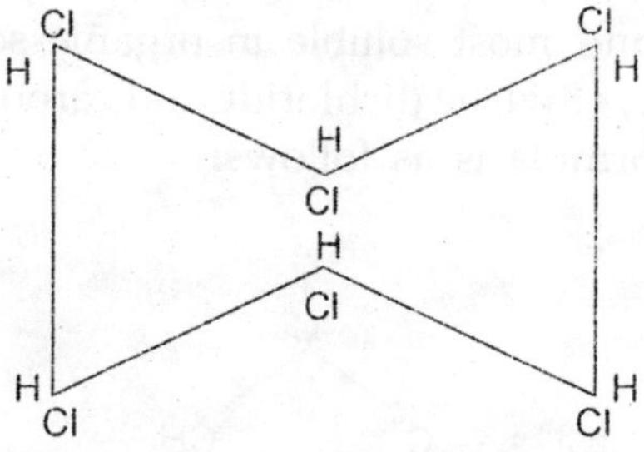

It is also known as gamma BHC as it contains 99% gamma isomer. The name lindane was proposed in 1949 after Vander Linden, a German Chemist who isolated this isomer in 1912. It is colourless crystals with a slight aeromatic odour. The lindane is practically insoluble in water (10 parts per million), slightly soluble in common paraphenic hydrocarbons and soluble in common organic solvents. It is a contact and stomach poison and has some fumigant properties. It is unstable in the presence of alkali and does not accumulate in the fat of animals. It can be used for controlling a wide range of vegetable pests on which BHC can not be used. The lindane is formulated as 20% emulsifiable concentrate, 10% granule and 1 to 3 % dust. Like BHC, it is also less poisonous to man and domesticated animals than DDT. Its oral and dermal LD 50 values to rat are 200 mg/kg and 500-1,000 mg/kg respectively.

Bicyclic Structure

These insecticides contain 2 non-aeromatic hydrocarbon rings fuse together so as to have 2 non-adjacent carbon atoms common to both rings. These are produced by the chlorination of the naturally occurring terpenes hence also called as chlorinated terpenes. It includes toxaphene and strobane.

Toxaphene ($C_{10}H_{10}Cl_8$)

This insecticide is a chlorinated camphene containing 67 to 69% chlorine. A technical product is a yellowish waxy solid melting point 65 to 90°C and is a mixture of isomers. It is unstable in the presence of alkali and upon prolonged exposure to sunlight. It is insoluble in water, readily soluble in paraphenic hydrocarbons and most soluble in organic solvents such as benzene, xylene, ethylene dichloride and carbon tetra chloride. Its structural formula is as follows:

```
                 H
                 |
                 C
               /   \
(CH3)2 — C           CH2
         |     HCH    |—— CH2
         |            |
  H2C =  C           CH2
               \   /
                 C
                 |
                 H
```

Toxaphene primarily acts as a stomach poison but some times is also used as contact poison. In comparison to BHC and DDT, it is slightly slow in action and has a lesser residual toxicity. It is non-toxic to honey bees but to human beings, it is four times more toxic than DDT. Toxaphene shows pytotoxicity to cucurbits and is also poisonous to fish. It is formulated as follows:

1. Dusting powder—It is available 5, 10 and 20%.
2. Water dispersible powder—25 to 40%.
3. Emulsifiable concentrate —Available as 25%.

Strobane ($C_{10}H_{11}Cl_7$)

It is prepared by chlorinating a mixture of camphene and pinene and containing 66% chlorine. The technical material is straw coloured liquid. It is practically insoluble in water but soluble in most organic solvents. It is not more stable to alkali. It has strong insecticidal and some acaricidal properties. Its oral and dermal LD 50 values, respectively, are 200 to 250 mg/kg for rat and 500 mg/kg for rabbit.

Polycyclic Structure

These compounds contain 3 or 4 non-aeromatic hydrocarbon rings with atleast two points of fusion. The insecticides of this group have been derived from a hexa chlorocyclo pentadiene and some times also called as the cyclodiene insecticides which also include the some insecticides of the next group (Heterocyclic structure). It includes following three insecticides.

Aldrin ($C_{12}H_8Cl_6$)

It is chemically 1, 2, 3, 4, 10, 10- hexachloro- 1, 4, 4a, 5, 8, 8a- hexahydro- 1, 4, -endo-exo-5, 8- dimethanonaphthalene. Aldrin is white crystalline compound with melting point 104°C, and containing 78% active aldrin. It is soluble in organic solvents such as acetone, amyl acetate, benzene, carbon tetrachloride, ethyl alcohol and xylene etc. Its structural formula is as follows:

Cl H
C H C
C
Cl—C HC
CCl_2 CH_2
Cl=C HC
C
C H C
Cl H

Aldrin is stable in strong alkali and is compatible with all commonly used insecticides. It is non phytotoxic to plant under most conditions of uses and free from smell. Aldrin is more effective insecticide than other chlorinated insecticides and has proved particularly valuable for the control of soil borne pests (cutworms, white grubs, termites and grasshoppers etc.) because of its prolonged residual action. It mainly acts as a contact or stomach poison and is more toxic than DDT. Its oral and dermal LD 50 values for rat are 40 to 50 and 200 mg/kg, respectively. Crops meant for human or cattle consumption should not be sprayed or dusted with aldrin three weeks prior to harvesting. It is formulated as under:

(i) *Dust:* Usually 5% dusting powder is available and used at the rate of 25-30 kg/hect.

(ii) *Wettable Powder:* Water dispersible powder of 40% is formulated which is applied at a concentration of 0.15 to 0.3%.

(iii) *Emulsifiable Concentrate:* It is available as 30% and used in the same concentration as wettable powder.

Note: Aldrin 5% dust is generally used for the control of cut-worms attacking potato crop and is applied in the soil at the time of sowing. Recently, the experiments conducted at Indian Agriculture Research Institute, New Delhi, revealed that the residues of aldrin were present in the tubers 10 times more than the permissible limit, besides it was also present in the form of dieldrin. Therefore, the aldrin should not be used in tuber crops.

Chlordane ($C_{10}H_6Cl_8$)

Chlordane or Octachlor was discovered by Julious Hyman in 1945. It is 1, 2, 3, 4, 5, 6, 7, 10, 10-octachloro-4, 7, 8, 9-tetrahydro-4, 7, methyleneindane. The technical material is a brown viscous liquid and has odour resembling cedar. Of the two isomers of chlordane β isomer is more toxic than α isomer. The technical material is unstable in the presence of alkaline material liberating hydrogen chloride and forming non-toxic product. Its specific gravity is 1.61. It is insoluble in water but highly soluble in common solvents. Chlordane is slowly volatile and, therefore, has a slight fumigant action besides being a contact and stomach poison. Its structural formula is as follows:

Chlordane should not be used on leafy vegetables or on root crops since it is likely to taint them. At higher concentrations it is toxic to cucurbits. It is also equally toxic to warm blooded animals as DDT and hazardous to fish. Its oral and dermal LD 50 values for rat are 283 and 1600 mg/kg respectively. It is formulated as under:

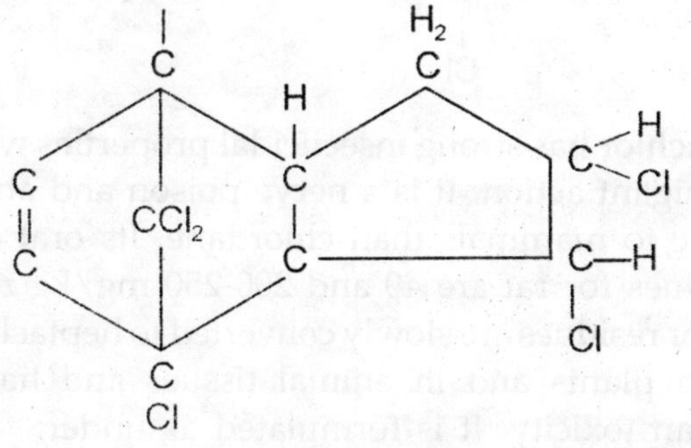

(i) *Dust:* Dusting powder is generally available as 5% and is used at the rate of 25 to 30 kg per hectare to control the grasshoppers and termites.

(ii) *Emulsifiable concentrate:* It is formulated as 20% and used in the strength of 0.15 to 0.30%.

(iii) *Wettable powder:* It is not in much use, previously it was formulated as 40 to 50% and used in the same strength as 'emulsifiable concentrate.

Heptachlor ($C_{10}H_5Cl_7$)

Heptachlor is 1, 4, 5, 6, 7, 10,10 heptachloro-4, 7, 8, 9-tetrahydro-4, 7-methylenein-dane. The technical product contains about 67% heptachlor and is waxy solid. The purified material is white crystalline solid with melting point 95-96°C. It is practically insoluble in water but variously soluble in acetone, benzene, CCl_4, ethyl alcohol, deodorised kerosine and xylene etc. It is stable to air, light, heat, moisture and mild alkaline conditions. Its structural formula is as follows:

```
          Cl
          |              H   Cl
          C                \ /
                            C
                   H
                   |
  ClC              C             CH
  ||      CCl2     |             ||
  ClC              C ----------- CH
                   H
          C
          |
          Cl
```

Heptachlor has strong insecticidal properties with possibly some fumigant action. It is a nerve poison and about 2 times more toxic to mammals than chlordane. Its oral and dermal LD 50 values for rat are 40 and 200-250 mg/kg respectively. Heptachlor residues are slowly converted to heptachlor epoxide in and on plants and in animal tissues and has the same mammalian toxicity. It is formulated as under:

(i) *Dust:* The dusting powder is available in the strength of 3-5% and used @ 25 to 30 kg per hectare to control the soil inhabiting pests as cutworms, wireworms, white grubs and termites etc.

(ii) *Emulsifiable Concentrate:* It is formulated in the strength of 20% and used as a spray in the concentrations ranging from 0.10 to 0.30%.

(iii) *Wettable Powder:* Available as 25% and used in the same concentration as emulsifiable concentrate.

Heterocyclic Structure

These compound consist of 1 or more rings as in the polycyclic structure but atleast one of the atoms of the ring is not a carbon atom, *i.e.,* it is usually N, O or S. It includes following insecticides:

Dieldrin ($Cl_2H_8Cl_6O$)

This is the epoxide of Aldrin and chemically it is 1, 2, 3, 4, 10, 10-hexa chloro-6, 7-epoxy-14a, 5, 6, 7, 8a-octahydro-l, 4-endo, exo-5, 8 dimethano naphthalene. The technical material is containing about 76% of active ingredient. The technical material is light tan flaky solid and when purified it exhists as white odourless crystals melting at 176°C. It is insoluble in water but moderately soluble in organic solvents. Its structural formula is as follows:

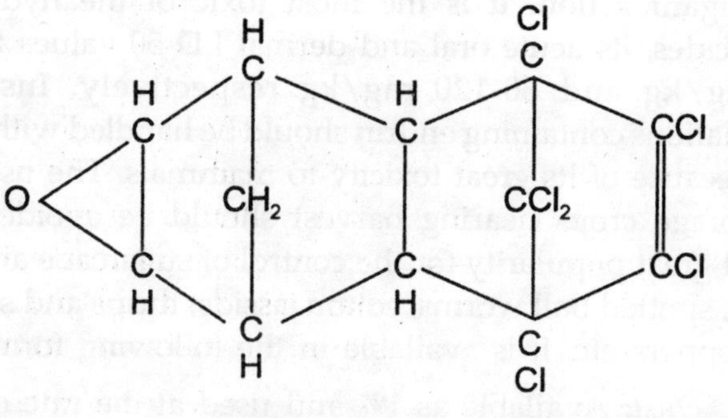

H Cl
C C
H H
C C CCl
O CH_2 CCl_2
C C CCl
H H
C C
H Cl

Dieldrin is very stable material and is not affected by alkali, light and mild acids. It has long residual action and is compatible with most of the insecticides and fungicides. It acts as a stomach and contact poisons and 10 times more toxic than Aldrin. Its oral and dermal LD 50 values to rat are 40 and 100 mg/kg, respectively. It is available in the following formulations and used to control the chilli thrips, termites, ants, grasshoppers, cutworms etc.

(i) *Dust:* 1.5% dusting powder is available which is used @ 25 to 30 kg/hect.

(ii) *Wettable Powder:* It is formulated in the strength of 20 to 50% and used in the concentration of 0.05 to 0.12%.

(iii) *Emulsifiable Concentrate:* It is available in the strength of 20% and used in the same concentrations as above.

Endrin **($C_6H_8Cl_6O$)**

It is 1, 2, 3, 4, 10, 10-hexa chloro 6, 7-epoxy-l, 4, 4a, 5, 6, 6, 8, 8a-Octahydro-exo-l, 4-endo, endo-5, 8 dimethamonaphthalene or the endo, endo isomer of dieldrin. The technical endrin is less stable than dieldrin and its formulations is stabilized by the addition of small amount of hexa methylene tetramine. It is insoluble in water, slightly in alcohols and petroleum oils and moderately soluble in acetone and benzene. Its structural formula is similar to dieldrin.

Endrin has a wide range of spectrum and high residual toxicity. It acts as a contact and stomach poison and also has a fumigant action. It is the most toxic of the hydrocarbon insecticides, its acute oral and dermal LD 50 values to rat are 3-6 mg/kg and 60-120 mg/kg respectively. Insecticidal formulations containing endrin should be handled with extreme care because of its great toxicity to mammals. The use in food and forage crops nearing harvest should be avoided. It has gained great popularity for the control of sugarcane and paddy borers, spotted boll worms, cotton jassids, thrips and sugarcane leaf hoppers etc. It is available in the following formulations:

(i) *Dust:* Available as 1% and used at the rate of 25 kg/hect.

(ii) *Emulsifiable Concentrate:* It is formulated as 20% emulsifiable concentrate which is used in the concentrations of 0.01 to 0.03% for the control of a wide range of pests in India.

(iii) *Granule:* Endrin 2% granule formulation is available which has been to be effective for the control of Lepidopterous borers infesting sugarcane, sorghum and rice etc.

Note: The government of India has imposed restriction since 1st April, 1977 in the use of endrin due to its high mammalian toxicity.

Thiodan ($C_9H_6Cl_6O_2S$)

This insecticide commonly known as endosulfan and contains both S and O, as heterocyclic atoms. Chemically it is 6, 7, 8, 9, 10, 10-hexachloro- 1, 5, 5a, 6, 9, 9a-hexahydro-6, 9-methano-2, 4, 3-benzo dioxathiepin 3-oxide. The technical material is brownish crystalline solid melting from 70-100°C. It is both a chlorinated hydrocarbon and an organic sulphite. The structural formula is as follows:

```
            Cl
            |
            C              CH2O
           /    H              \
ClC              C              \
 ||     CCl2     |               SO
ClC              C              /
           \    H              /
            C              CH2O
            |
            Cl
```

It is practically insoluble in water but moderately soluble in common organic solvents. It acts as contact and stomach poison and also has slight fumigant action. Thiodan is general purpose insecticide effective against many different types of insect pests. It is not harmful to bees and natural enemies of pests. Its acute oral and dermal LD 50 values of rat are 110 and 74 to 130 mg/kg respectively. It is specially effective against borers, aphids, leaf hoppers and other Homopterous pests. It is available in the following formulations:

(i) *Dust:* 4% dust generally available which is used @ 25-30 kg/hect. It can be safely used on cucurbits to control the red pumpkin and epilachna beetles.

(ii) *Emulsifiable Concentrate:* It is available in the strength of 35% and used in the concentration of 0.05 to 0.07%.

(iii) *Granule:* 4% granules are available and used for the control of soil inhabiting pests and stem borers of sorghum and sugarcane.

Organo-phosphorus Insecticides: The insecticides of this group contain one or more atoms of phosphorus bonded either directly to the carbon atoms or indirectly through N_2, O_2 or S. Schradar and his associates just prior to IInd world war discovered the insecticidal potency of this chemical group in Germany. The insecticides of this group mainly act as contact and stomach poisons, and some are systemic, some slowly give out vapours which have fumigant action. All the organophosphorus insecticides appear to have a common mode of action.

They are essentially irreversibly inhibitors of cholinestrase enzyme of the neuro-muscular system. They have the decided advantage of being detoxify very rapidly to animal tissues and eliminated rather than these stored in fatty tissues as most of the chlorinated hydrocarbon insecticides.

The insecticides of this group are classified on the basis of phosphorus atoms found in the combination as phosphatase (TEPP, DDVP, chlorfenvinphos, tetrachlorvinphos, phos-phamidorr, monocrotophos, dicrotophos and mevinphos), phosphonates (trichlorphon), phosphorothionates (parathion, fenitrothion, fenthion fensulfothion, chlorpyriphos, thionazin, diazinon, methyl-demeton, demeton and quinalphos), phosphorothioiates + (oxydemeton-methyl) phosphorothiolothionates (ethion, phorate, thiometon, disulfoton, dimethoate, formothion malathion, phenthoate, carbophenothion and phosalone), phosphonothionates (EPN and leptophos), Phosphoramidothioate (methamidophos), phosphoramidate (schradan, mephosfolan and phenamiphos).

On the basis of their mode of action these insecticides may be classified into two groups.

Contact Insecticides

It includes following chemicals:

1. HETP (*Hera ethyl tetra phosphate)*
2. Sulphotate
3. Parathion
4. Ethyl parathion
5. Methyl parathion
6. Malathion
7. Diazinon
8. Guthion
9. Trithion
10. Ethion
11. Dipterex
12. DDVP
13. Dibrom
14. Dursban
15. Zolone
16. Oftanol
17. Volaton
18. Hostathion
19. Tamaron
20. Elsan

Systemic Insecticides

These are those chemicals which are absorbed by the plant system and translocated in the whole parts of the plants. The some of the important insecticides are as follows:

1. Shradan
2. Dimefox
3. Demeton
4. Methyl-o-demeton
5. Rogor
6. Phosdrin
7. Dimecron
8. Phorate
9. Disyston
10. Ekatin
11. Dasanit
12. Azodrin
13. Cytrolane
14. Formothion
15. Phosvel

The important and commonly used insecticides are described below:

Malathion ($C_{10}H_{19}O_6PS_2$)

The technical material is 95% to 98% pure with an unpleasant odour. The purified compound is brownish liquid and chemically it is 0, 0-dimethyls (1, 2-dicarbe thoxyethyl) Phosphorodithioate. It is soluble in water to only 0.0145%, slightly soluble in mineral oils and soluble in most organic solvents.

It is unstable in alkaline medium. Its structural formula is as follows:

$$\begin{matrix} CH_3O \\ CH_3O \end{matrix} \Big> \overset{S}{\overset{\|}{P}}S \;\; \begin{matrix} CH-\overset{O}{\overset{\|}{C}}O\;C_2\;H_5 \\ | \\ CH_2-\underset{\underset{O}{\|}}{C}O\;C_2\;H_5 \end{matrix}$$

Malathion is mostly used as contact insecticide though it acts as a stomach poison and exhibits some fumigant action also. It is used for the control of a wide variety of pests of crops such as aphids, hoppers, psyllids, bugs, caterpillars, floora-beetles and mites etc. The malathion is the safest insecticide available so far, its acute oral and dermal LD 50 to rat being 1400 to 1900 and 4000 mg/kg respectively. At the recommended dosages it can be applied to most edible crops within one to three days of harvest without danger of excess harmful residues. It is formulated as dusts, wettable powders and emulsifiable concentrates and is being marketed as follows:

1. Dusting powder-5 per cent used 25 to 30 kg/hect.
2. Water dispersible powder-25 per cent and used in concentration of 0.025 to 0.03%.
3. Emulsifiable concentrate-50 per cent used in same concentration as above.
4. TJLV preparation: 96 per cent used in aerial spraying.

Diazinon ($C_{12}H_{21}N_2O_3PS$)

Chemically it is diethyl *o*-isopropyl-4 methyl-6-pyrimidinyl phosphorothionate and also known as Basudin. Diazinon is a colourless liquid with a slight characteristic odour and technical product is 90 per cent pure. It is slightly soluble in water (0.004%) and soluble in most of the organic solvents. Its structural formula is as follows:

```
                      C
                N         CH
CH3O                      ||          OC2   H5
      > CH—C              C—O—P <
CH3O                            ||    OC2   H5
                  N             S
```

Diazinon is a strong contact and stomach poison and has fumigant action and penetrating quality. Its acute oral and dermal LD 50 for rat are 150 to 220 and 500 to 2000 mg/kg respectively. It is used to control a variety of agricultural pests, *viz.*, lepidopterous larvae, bugs, thrips, beetles and mites etc. It has proved to be an outstanding insecticide for the control of most sucking insects, borers and leaf caterpillars. It is formulated as follows:

1. *Dust:* 2 to 5 per cent dusting powder and used @ 25 kg/hect.
2. *Emulsifiable Concentrate:* 25 per cent used in the concentration of 0.025 to 0.03 percent.
3. *Granule:* 5 to 10 per cent and used @ 25 kg/hect to control the soil inhabiting pests.

Trichlorfon ($C_4H_8CI_3O_4P$)

Chemically it is 0, 0 dimethyl 2, 2, 2-trichloro hydroxyethyl phosphonate. Purified chemical is a white crystalline solid, melting point 83-84°C. It is moderately soluble in water (15 per cent) and freely soluble in most organic solvents. Its structural formula is given below:

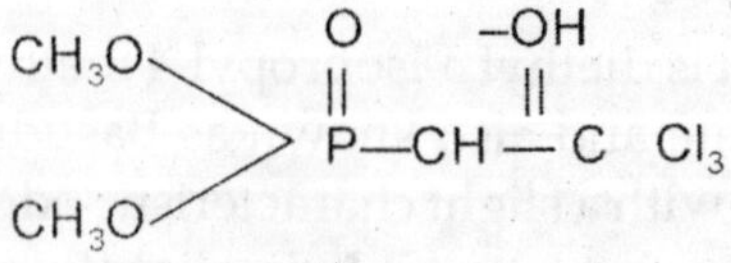

CH_3O O –OH
P—CH—C Cl_3
CH_3O

Trichlorfon is a contact and stomach poison with some fumigant action and has some penetrating effect also. It is a quick acting chemical effective against Lepidoptera, Diptera and Hemiptera. Under natural conditions it is readily dehydrochlorinated to dichlorvos. It has low mammalian toxicity but toxic to insects and effective up to 10 days of application. Its acute oral and dermal LD 50 values for rat are 560 and 72,800 mg/kg. Trichlorfon is also used in baits to control the fruit flies. It is generally available as 5% dust and 80 per cent wettable powder. Its trade names are dipterex; dylox and tugon.

Dichlorvos or DDVP ($C_4H_7O_4CI_2P$)

Chemically it is 2, 2 dichlorovinyl dimethyl phosphate. It is one of the products resulting from the alkaline treatment of trichlorfon. DDVP is thin colourless liquid with pleasant smell, boiling point 120°C. It is insoluble in water but readily soluble in most organic solvents. Its structural formula is as follows:

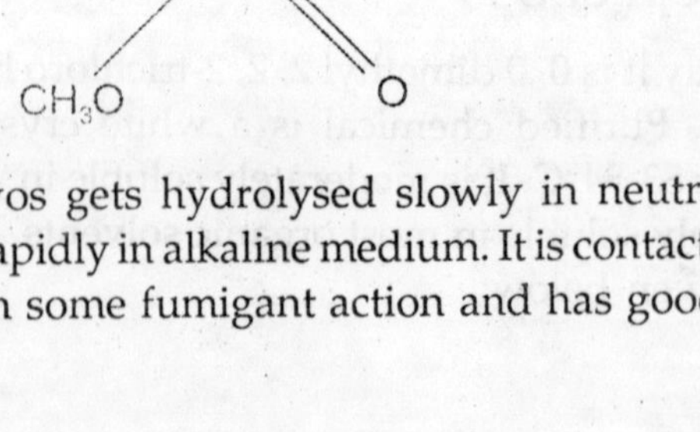

Cl
CH_3O OCH = C
Cl
P
CH_3O O

Dichlorvos gets hydrolysed slowly in neutral and acidic media and rapidly in alkaline medium. It is contact and stomach poisons with some fumigant action and has good penetration properties.

It brings about rapid knock down of the insects. It is highly volatile compound, therefore, can be used on all crops just before harvest. DDVP is very effective in baits and in aerosol formulations for the rapid knock down of flies, mosquitoes and moths etc. Its acute oral and dermal LD 50 values for rat are 25-30 and 75 to 900 mg/kg. It is available in the form of 78 EC with the trade names Nuvan and Vapona etc.

Carbophenothion

Chemically it is *o, o*-dimethyl-S para chlorophenyl thiomethyl phosphorodithioate. It is light amber colour liquid insoluble in water and soluble in common organic solvents including the paraphenic hydrocarbons. The pure chemical is a colourless, odourless oil. It hydrolyses very slowly in water suspension. Its structural formula is as follows:

$$(C_2H_5O)_2\,\overset{S}{\overset{\|}{P}}—S—CH_2—S—C_6H_4—Cl$$

It is a general purpose insecticide with long residual action in the field and effective against a wide range of insects and mites, particularly aphids, thrips, scale insects, leaf hoppers etc. Its acute oral and dermal LD 50 values to rat are 7-30 and 800 mg/kg respectively. It is formulated as 15 per cent EC and 25 per cent W.P. and sold in the names of Trithion and Garrathion.

Ethion ($C_2H_{22}O_4P_2S_4$)

Chemically it is 0, 0, 0, 0-tetraethyl S, S-methylene bis phosphorodithioate. Ethion is a colourless liquid insoluble in water but soluble in most organic solvents. It hydrolyses under both acidic and alkaline conditions.

Its structural formula is as follows:

$$(C_2H_5O)_2\,\overset{S}{\overset{\|}{P}}—S—CH_2—S—\overset{S}{\overset{\|}{P}}\,(OC_2H_5)_2$$

It is useful for the control of many insects, *viz.*, aphids, scale insects, leaf hoppers, leaf miners, thrips, bugs and mites. Effectiveness of ethion for control of scale insects and mites appear to be more by formulating it in petroleum oil. It is generally formulated as 50% EC, 25 to 50% W.P. and 2% solution in spray oil.

Chlorpyrifos ($C_9H_{11}CI_{13}NO_3PS$)

Chemically it is *o, o*-diethyl *o*- (3, 5, 6-trichloro-l-pyridyl) phosphorothioate. Chlorpyrifos is white crystalline substance with mild mercaptan odour. It is practically insoluble in water but readily soluble in most organic solvents. It is relatively stable to hydrolysis and oxidation. Its structural formula is as follows:

Cl Cl S
Cl —O—P O C_2H_5
N O C_2H_5

It is a contact and stomach poison with vapour action also and effective against many pests with particular reference to rice pests, borers, aphids, mosquitoes, soil inhabiting insects and some mites. It is formulated as 20 EC and available in the name of Dursban. Its acute oral and dermal LD 50 values to rat and rabit are 135 mg/kg and 1000 to 2000 mg/kg respectively.

Quinalphos ($C_{12}H_{15}O_3H_2PS$)

Chemically it is *o, o*-diethyl-o (quinoxalinyl, (2) thionophosphate. Quinalphos is an odourless white crystalline solid, melting point 31-32°C. It is practically insoluble in water but readily soluble in organic solvents. It is hydrolysed in a alkaline medium. Its structural formula is as follows:

N S O C_2H_5
P
O O C_2H_5
N

It is a strong contact and stomach poison and effective against a wide spectrum of insect pests of crops. Quinalphos is also used as effective acaricide. It is formulated as 25 per cent EC and 5 percent granule and available in the name of Ekalux. Its acute oral LD 50 for male rat is 137 mg/kg and dermal 1400 mg/kg.

Systemic Insecticides

Those insecticides which are translocated to all the parts of the plants when applied on the roots, stems, leaves or seeds, are known as systemic insecticides. The important and commonly used insecticides of this group are described here. These insecticides have following merits:

1. Equal coverage is possible.
2. Protecting new plant growth.
3. Having less harmful effects on beneficial, predatory and pollinating insects.
4. Increasing the residual effect as less affected by weathering.

Scharadan

It is octamethyl pyrophosphoramide and was among the 1st systemic insecticides. The technical material is brown liquid and contains 40 per cent of schradan boiling point 154°C. It is soluble in water and also remarkable stable to alkaline hydrolysis. Its structural formula is as follows:

$$[(CH_3)_2N]_2 - \overset{\overset{\displaystyle O}{\|}}{P} - O - \overset{\overset{\displaystyle O}{\|}}{P} - [N(CH_3)_2]_2$$

It possesses only a week contact action to most insect and converted slowly in plants and rapidly in animals to the toxic monoamide oxide. It is not phyto-toxic to most plant when used at the recommended doses. The schradan is specially effective against aphids and mites. Its acute oral and dermal LD 50 values to rat are 5 and 50-100 mg/kg respectively. It is also known as Pestox III. Systam and OMPA.

Demeton

It is mixture of 2 parts of 0, 0-diethyl 0-2 (ethylthio) ethyl phosphorothionate and I part of demeton-s-ie, diethyl 2-(ethyl thio) ethyl phosphorothionate. The technical material is yellowish liquid, soluble in organic solvents and unstable in alkaline solution. Both the isomers are only slightly soluble in water. Its structural formula is as follows:

$$\text{2 parts } (C_2H_5O)_2\overset{\displaystyle S}{\overset{\|}{P}}OCH_3\,CH_2\,SC_2H_5$$

$$\text{1 part } (C_2H_5O)_2\overset{\displaystyle O}{\overset{\|}{P}}SCH_2\,CH_2\,SC_2H_5$$

Demeton provides a long systemic action rapidly observed by roots, stems or foliage. It is particular effective as a spray for the control of many species of aphids and red spider mites and has no marked phyto-toxic effect. Its acute oral and dermal LD 50 values for rat are 3-5 and 200 mg/kg, respectively. It is formulated as 25 Ec and sold in the trade name of systox.

Demeton-o-methyl

It is 0, 0-dimethyl s-2-(ethyl-thio)-ethyl phosphorothionate. It has very similar property in behaviour as demeton but less toxic than the former. Its acute oral and dermal LD 50 values for rat are 50 to 75 and 300 to 450 mg/kg respectively. Its structural formula is as follows:

$$\begin{matrix} CH_3O \\ CH_3O \end{matrix} > \overset{\displaystyle O}{\overset{\|}{P}}SCH_2\,CH_2\,SC_2H_5$$

It is effective against sucking insect pests and mites of crops. It is formulated as 25 per cent EC and available in the name of Metasystox.

Dimethoate ($C_5H_{12}O_3PS_2N$)

Chemically it is *o*, *o*-dimethyl S (N-methyl carbomoylmethyl) Phosphorodithioate. Dimethoate is a white crystalline solid, melting point 121.8 to 123.6°F and technical material contains 94 to 96 per cent chemical. It is soluble in organic solvents and also in water to the extent of about 7% and unstable in alkaline media. Its structural formula is as follows:

$$\begin{array}{l} CH_3O \diagdown \\ \qquad\quad P - S\ CH_2\ CO\ NH\ CH_3 \\ CH_3O \diagup \| \\ \qquad\qquad S \end{array}$$

It is a systemic and contact poison effective against sucking insect pests and mites of a wide variety of crops and fruit trees. It is highly toxic to insects but less toxic to mammals, its acute oral and dermal LD 50 values for rat are 200 to 300 and 700 to 1150 mg/kg respectively. It is generally formulated as 30 per cent EC and sold in the name of Rogor.

Mevinphos

Chemically it is dimethyl 2-carbomethoxy-l-methylvenyl phosphate. The technical material is colourless liquid, melting point 106-107°C miscible in water and soluble in most of the organic solvents. The technical product is composed of about 2 parts of *trans*-isomer and I part of *cis*-isomer. The *trans*-isomer is about 10 times as active as *cis*-isomer.

Its structural formula is as follows:

$$\begin{array}{l} \qquad\qquad O \qquad\qquad\qquad\qquad O \\ CH_3O \diagdown \| \qquad\qquad\qquad\qquad \| \\ \qquad\qquad POC\ (CH_3) = OH\ CoH_3 \\ CH_3O \diagup \end{array}$$

It is a systemic insecticide and has powerful contact and fumigant action. It is rapidly broken down in plants, therefore, its residues are very short lived in the field and is specially useful to control the pests on food crops close to harvest. It is highly poisonous by injection, absorbtion through skin and inhalation.

Areas inhabited by wild life and fish should not be contaminated with this. Its acute oral and dermal LD 50 values to rat are 3 to 5 and 90 mg/kg respectively. It is formulated as 25 per cent EC and 10 per cent W.P. and sold in the name of phosdrin.

Phosphamidon ($C_{10}H_{19}O_5NClP$)

Chemically it is *o, o*-dimethyl-o- (I-methyl-2-chloro-2-diethyl-carbamoyl-vinyl) phosphate. It is a colourless liquid, boiling point 160°C with faint pleasant odour. Technical grade is dark brown and commercial product is bright violet due to addition of a dye. It is miscible with water and all organic solvents except saturated hydrocarbons in which it is soluble only to a limited extent.

Its structural formula is as follows:

$$(CH_3O)_2P(=O){-}O{-}C(CH_3){=}C(Cl){-}CO{-}N(C_2H_5)_2$$

Phosphamidon is a systemic insecticide and acts as a stomach poison and has a relatively low contact action. It is rapidly absorbed by the plant surface and is quickly decomposed in the plant. Metabolism of phosphamidon gives rise to desethyl phosphamidon which is roughly as toxic as the present compound. It is effective against many sucking, chewing and mining insects and certain mites. Its acute oral and dermal LD 50 values for rat are 15 and 125 mg/kg respectively. It is formulated as 85% and available in the trade name of Dimecron.

Monocrotophos ($C_7H_{14}O_5NP$)

Chemically it is dimethyl *cis*-l-methyl-2-methyl carbamoyvinyl phosphate. It is made up of colourless crystals with mild ester smell, boiling point 125°C. Monocrotophos is miscible with water and all organic solvents. It is rapidly hydrolysed in alkaline solutions than in acids. Its structural formula is as follows:

$$\begin{array}{l} CH_3O \\ \quad\;\; \rangle P(=O)-O-C(CH_3)=CH-C(=O)-N(H)(CH_3) \\ CH_3O \end{array}$$

Monocrotophos is a systemic insecticide but also acts as a stomach and contact poison. It is a hazard to wild life if the food is contaminated with residues above 0.5 ppm. Its acute oral LD 50 value for rat is 20 mg/kg. It is effective against a wide variety of insect pests, *viz.*, thrips, aphids, leaf miners, cotton boll worm, white flies and some mites. It is generally available as 40% SC and sold in the name of Azodrin and Nuvacron.

Formothion ($C_6H_{12}O_4PS_2N$)

Chemically it is *o*, *o*-dimethyl-S-(N-methyl-N-formyl) carbamoyl-methyl-dithiophosphate. It is yellowish viscous oil or crystalline solid with a garlic like odour, melting point 25-26°C. Its structural formula is as follows:

$$\begin{array}{l} CH_3O \\ \quad\;\; \rangle P(=S)-S-CH_2-C(=O)-N(CH_3)(CHO) \\ CH_3O \end{array}$$

It is a systemic insecticide and acaricide with contact action. Formothion is absorbed by the plant and hydrolysed into such metabolites which are less harmful to mammals but toxic to insect pests. Its acute oral and dermal LD 50 values to rat are 400 and 400-1680 mg/kg respectively. It is formulated as 25% EC and available in the trade name of Anthio.

Thiometon ($C_6H_{15}O_2PS_3$)

Chemically it is O, O-dimethyl S-ethyl mercaptoethyl-dithiophosphate. Thiometon is a dithioate analogue of methyl-O-demeton. Its structural formula is as follows:

$$\begin{array}{l} CH_3O \\ \quad\quad\quad \overset{S}{\overset{\|}{P}} - S - CH_3 - CH_2 - S - C_2H_5 \\ CH_2O \end{array}$$

It is a strong systemic insecticide and acaricide effective against sucking insects with particular reference to aphids and sawflies. It is comparatively safer to beneficial and predatory and parasitic insects. It is hydrolysed slowly in plant system by oxidation and saponification and remain effective upto 2-3 weeks. Its acute oral and dermal LD 50 values to rat are 100 and 200 mg/kg respectively. Thiometon is formulated as 25% EC and sold in the trade name of Ekatin.

Phorate ($C_7H_{17}O_2PS_3$)

Chemically it is *o*, *o*-diethyl-i-(ethylthio) methyl phosphorodithionate. It is a yellowish liquid, boiling point 118-120°C. Phorate is slightly soluble in water but is miscible with many organic solvents. It gets hydrolysed in the presence of moisture and alkaline conditions. Its structural formula is as follows:

$$\begin{array}{l} C_2H_5O \\ \quad\quad\quad \overset{S}{\overset{\|}{P}} - S - CH_2 - CH_2 - S - C_2H_3 \\ C_2H_5O \end{array}$$

It is systemic insecticide and is absorbed by roots and subsequently translocated to aerial parts of the plant. This material some times causes reduction in stems from seed treatment. It has also shown some nematicidal and acaricidal action. It has been found useful in the control of sorghum shoot fly, aphids, sucking pests of cotton, white grub and gall midge of rice etc. Its acute oral and dermal LD 50 values to rat are 2 to 3 and 70 to 300 mg/kg respectively. It is generally formulated as 10% granule and available in the trade name of Thimet.

Thiodemeton or Disulfoton ($C_8H_{19}O_2PS_3$)

Chemically it is *o, o*-diethyl-S-2-(ethylthio) ethyl phosphorodithioate. It is a colourless oily liquid with characteristic odour, Boiling point 62°C. Disulfoton is slightly soluble in water and soluble in most of the organic solvents. Its structural formula is as follows:

$$\begin{array}{l} C_2H_5O \diagdown \quad \overset{S}{\overset{\|}{P}} - S - CH_2 - CH_2 - S - C_2H_3 \\ C_2H_5O \diagup \end{array}$$

It is oxidised in plant tissues to form the sulfoxide and sulfone metabolities which are responsible for the systemic toxicity of the compound.

Sufficient moisture should be present in the soil for it to act effectively. It is specially effective for preplanting or seed dressing application as a seed treatment to control insects and mites after emergence of the plants. Its acute oral and dermal LD 50 values for rat are 4 and 50 mg/kg respectively. It is formulated as 5% granule and sold in the trade names of Disyston and Solvirex.

Carbonates: The carbamate insecticides are comparatively recent in development and are the derivatives of carbamid acid and have an OCOn = group in the molecule. They chiefly act by inhibiting the cholinestrase enzyme of the neuromuscular system. The commonly used insecticides of this group are as follows:

Carbaryl

It is I-naphthyl-N-methyl-carbamate and available in the name of Sevin. Carbaryl is white crystalline solid, m.p. 142°C.

The technical material is 95% pure. It is very slightly soluble (0.004%) in water and moderately soluble in most organic solvents. It has the following structural formula:

O
||
$OCNHCH_3$

It is a contact and stomach poison and has a longer residual action. Carbaryl is only occasionally weakly phytotoxic particularly on apple foliage. Excessive doses are likely to reduce seed germination. It is general purpose insecticide and has been found specially effective for fruit flies and borer pest. Its acute oral and dermal LD 50 to rat are 400 and 7500 mg/kg respectively. It is formulated as 5% to 10% dusts and 50% wettable powder. Sevimol 40 LV is a product containing 40% carbaryl with molasses. Granule containing 4% in each of carbaryl and gamma BHC is formulated with the name of Sevidol.

Isolan

It is a product containing l-isopropyl-3-methyl-5-pyrazolyl dimethyl-carbamate. Isolan is a colourless liquid boiling point 105°C to 107°C. It is miscible with water and soluble in most organic solvents. Its structural formula is as follows:

CH_3 C— CH O
|| || ||
N CO CN $((CH_3)_2$
N
$CH(CH_3)_2$

Isolan is a systemic insecticide and also acts as contact, stomach and fumigants. It is mainly used against aphids and in fly baits. Its acute oral and dermal LD 50 values to rat are 12 and 35-60 mg/kg, respectively.

Aminocarb

It is the common name for the product Metacil having 3-methyl-4-dimethylanimophenyl-N-monomethyl carbamate. Aminocarb is a white crystalline odourless substance, m. p. 93-94°C. It is slightly soluble in water, moderately soluble in aromatic solvents and readily soluble inorganic solvents. It is unstable in strongly alkaline medium. Its empirical formula is $C_{11}H_{16}O_2N_2$ and structural formula is as follows:

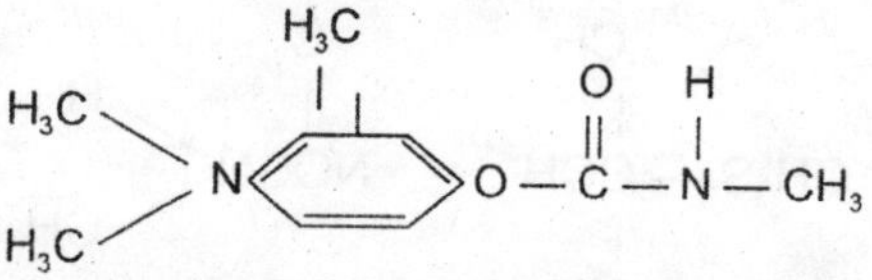

Aminocarb is effective against biting insects with particular reference to Lepidop-terous larvae and also some acaricidal properties. Its acute oral and dermal LD 50 values to rat are 30-50 and 275 to 1000 mg/kg respectively. It is generally formulated as 5% dust and 50% wettable powder.

Carbofuran

Chemically it is 2, 3-dihydro-2-2dimethyl-7-benzofuranyl methylcarbamate. It is a white crystalline solid with slightly phenolic odour, m.p. 153-154°C . Carbofuran is slightly soluble in water (250 to 700 K ppm) and soluble in most organic solvents. It is non-inflammable and unstable inalkaline media. Its empirical formula is $C_{12}H_{15}O_3NO_3$ and has the following structural formula:

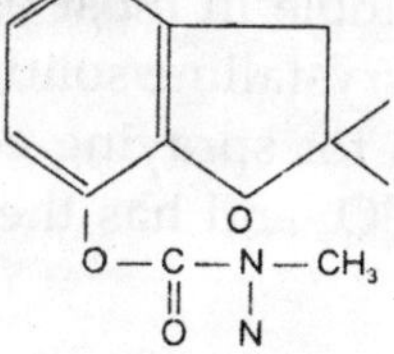

Carbofuran is systemic insecticide and effective against sucking pests, thrips, mites, nematodes and other soil inhabiting pests. Its acute oral and dermal LD 50 values to rat are 8.2-14.1 and 10,200 mg/kg respectively. Its trade name is Furadan and formulated as 3% granule and 50% S.P. or F.P.

Aldicarb

Chemically it is 2-methyl-2-(methyl thio) propionaldehyde 0-l-(methyl carbamoyl) oxime. Aldicarb is white crystalline solid with slightly sulfurous odour, m.p. 98-100°C. It is slightly soluble in water (0.6%) and more soluble in organic solvents. Its empirical formula is $C_7H_{14}N_2O_2S$ and has the following structural formula.

```
          CH3           C
          |             ||        H
CH3S --- CCH    =     NOCN <
          |                       CH3
          CH3
```

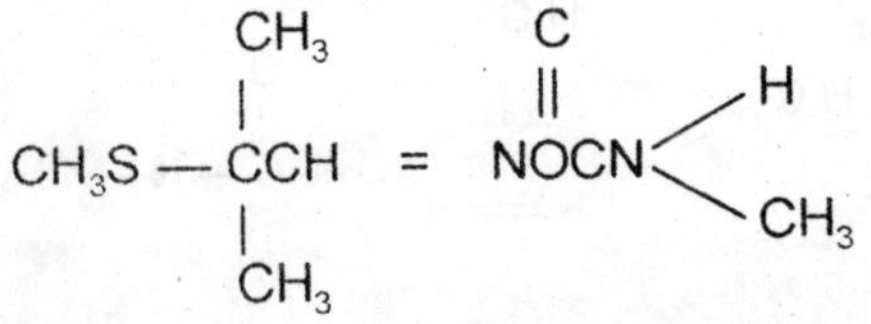

Aldicarb is a systemic insecticide, acaricide and nematicide, effective against a wide variety of insects, mites and nematodes. After application in the soil, it is rapidly absorbed by the plants and oxidised to the principal toxicant aldicarb sulfoxide which is further oxidised to the less toxic aldicarb sulfone. Its acute ora and dermal LD 50 values to rat are 6.2 and 3200 mg/kg respectively. It is formulated as 10% granule and available under the trade name of Temik.

Primicarb

Chemically it is 2-dimethylamino-5, 6 dimethylpyrimidinyl dimethyl carbamate. It is fast acting selective aphicide of moderate mammalian toxicity. Primicarb is not miscible with water but very soluble in most organic solvents, m.p. 90.5°C. It is colourless, crystalline solid, non-corrosive to the material normally used for spraying equipment. Its empirical formula is $C_{11}H_{18}N_4O_2$ and has the following structural formula.

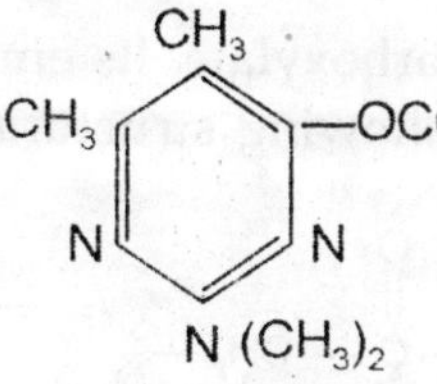

$$\text{pyrimidine ring: } CH_3 \text{ (top)},\ CH_3 \text{ (left)},\ -OCON(CH_3)_2 \text{ (right)},\ N,\ N,\ N(CH_3)_2 \text{ (bottom)}$$

It is a contact selective aphicide, has a rapid knock down and high activity at low concentrations and also act as a fumigant with no unpleasant odour. Its acute oral LD 50 value to rat is 147 mg/kg. It is formulated as 50% dispersible powder and available under the trade name Pirimor.

Synthetic Pyrethroids

The synthetic pyrethroids are mostly based on the earlier natural and synthetic pyrethroids, but differ so markedly from them in properties and action so as to constitute a new class of insecticides. The first synthetic pyrethorid was described in 1973 by *Elliott* and co-workers. The structure of the natural pyrethrins, of which pyrethrin I is the prototype, has been modified to give successive synthetic compounds with improved properties. The following are the characteristics of these insecticides:

1. They are quick in action.
2. These have high insecticidal efficiency.
3. Comparatively have low mammalian toxicity.
4. Good persistence.
5. Amenability to synergistic action.
6. Small quantity needed to kill the insects.
7. Less risk of environmental contamination.
8. Less chances of insecticidal resistance.

The important insecticides of this group are as follows:

Permethrin

It is also known as ambush and PP 557. Chemically it is 3-phenoxybenzyl (±) *cis, trans* 2-(2, 2-dichlorovinyl)-2, 2-

dimethyl cyclopropane-1 carboxylate. Its empirical formula is $C_{21}H_{20}O_3C_{12}$ and has the following structural formula:

```
                                  O
Cl                                ||
   \ C = CH — CH — CH — C — CH2 —(benzene ring)
Cl /            \   /                       \
                  C                          O —(benzene ring)
                /   \
             CH3     CH3
```

It is practically insoluble (0.2 ppm) in water but miscible with most organic solvents. The compound is both a stomach and a contact insecticide. It is extremely fast acting and has shown adulticidal, ovicidal and particularly larvaecidal activity. Thus it possesses a very high level of activity against a wide range of insect pests, particularly Lepidoptera. Its acute oral and dermal LD 50 value to rat is 74000 mg/kg. It is formulated as 25 and 50% emulsifiable concentrate.

Cypermethrin

Also known as cymbush and PP 383. Chemically it is cyano-3-phenoxy benzyl (±) *cis. trans*-3-(2, 2-dichlorovinyl)-2, 2-dimethyl cyclopropane-carboxylate. Its empirical formula is $C_{22}H_{19}O_3NCl_2$ and has the following structural formula:

```
                                  O       CH
Cl                                ||      |
   \ C = CH — CH — CH — C — O — CH2 —(benzene ring)
Cl /            \   /                          \
                  C                             O —(benzene ring)
                /   \
             CH3     CH3
```

It is practically insoluble in water (0.2 ppm) but miscible with most organic solvents. It possesses a very high level of activity against a wide range of insect pests, particularly Lepidoptera. The activity is usually 10 times greater than that of organophosphates and carbonates. Unlike natural pyrethrins, it is stable to degradation in light, hence, it is relatively persistent on plant surfaces, even under extremes of temperature and sunlight. Its acute oral and dermal LD 50 values to rat are 4030 and 4800 mg/kg respectively. It is formulated as 25% emulsifiable concentrate.

Decamethrin: It is also known as decis. It acts on insects by contact and has a direct insecticidal effect on biologically active instars of pests. It is not broken down rapidly by the enzymes in the insects, and larvae die within a few hours. It is 100, 80, 50 and 40 times more powerful than DDT, carbaryl, malathion and parathion, respectively. Decamethrin shows resistance to leaching out during heavy rain or sprinkler irrigation. The acute oral toxicity to rat is (LD 50) 128.5 to 138.7 mg/kg. It is formulated as 2.8 w/w emulsifiable concentrate containing 25 decamethrin per litre.

Fenvalerate: It is also known as sumicidin and chemically it is cyano-*m*-phenoxy benzyl-isopropyl-*p*-chlorophenyl-acetate. Its empirical formula is $C_{25}H_{22}ClNO_3$ and has the following structural formula:

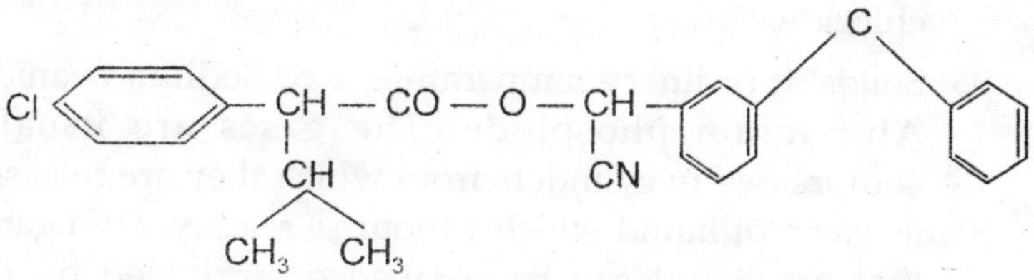

Cl — C₆H₄ — CH — CO — O — CH — C₆H₄ — O — C₆H₅

(CH below first CH bearing CH_3 CH_3; CN below second CH)

Fenvalerate is a yellowish oily liquid. It is not miscible with water (< 20 ppm) but soluble in most organic solvents. Its insecticidal spectrum is relatively wide and specially effective against Lepidoptera, Hemiptera and Diptera larvae. It is photostable, therefore, its efficacy lasts long in the field. It has quick killing and knockdown effects. The acute oral and dermal LD 50 values to rat are 541 and 73200 mg/kg, respectively. It is formulated as 20% emulsifiable concentrate.

Fumigants

Gaseous poisons use to kill insects are called as fumigants. The application is generally restricted to plants or products in tight enclosures or to soil. They may be used to control all types of insects with irrespective of types of mouth parts since the gas enters the insect body through the spiracles during respiration.

The fumigant which vapourises readily at room temperature is the most useful. For soil fumigation, however, the slower release of vapour from substances, boiling at high temperature has proved most effective. An ideal fumigant is selected by its safety to human beings, animal and plants; penetration quality; non inflammable; harmless to food stuff; non-corrosive to metals; have great power of diffusion; cost; effect on germinating seeds and relative effectiveness. The fumigants may be divided into three broad groups:

1. Gases at ordinary temperature, *e.g.*, Methyl bromide, Ethylene oxide etc.
2. Liquids at ordinary temperature, *e.g.*, Ethylene dichloride, Ethylene dibromide and Carbon tetra chloride.
3. Solids at ordinary temperature, *e.g.*, Sodium cyanide; Aluminium phosphide. The gases are usually compressed in cylinders from which they are released as gases of liquid which vapourise readily. Fumigants that are liquid can be sprayed or sprinkled on the commodities or in the enclosed space soil fumigants can be vapourised by heating or simply by mixing them in the commodity. In the operation of any fumigant the first principle to be observed is to safeguard human lives and only trained personnel with necessary protective appliance should be entrusted with the work. The space to be fumigated must be nearly airtight and it is better to avoid windy or cold weather. There should be provision for ventilation after fumigation.

The dosage for fumigation is expressed in terms of kg/ 1000 cu.m. In fumigation, important things is the amount of gas acting on the insects over a certain period of time. Successful fumigation is dependent upon the attainment of a critical value (ct) which will kill atleast 99% of the insect population. The ct value is the product of the concentration of the gas per unit volume x the duration of the exposure (time). Therefore, it is necessary to know the volume of the enclosed space to be treated and the rate of release of the fumigant vapour. It is apparent that the longer the period of exposure, the lower the concentration of gas required. The sorption of the gas by the material fumigated affects the dosage and ct value of the fumigant, and further, the rate of sorption is dependent on the moisture content of the material. The commonly used fumigants are described here as under.

Hydrogen Cyanide or Hydro Cyanic Acid (HCN)

Of all the chemicals used as fumigants for insects, HCN is extremely used inspite of its extreme toxicity to the human beings. It was originally employed for the control of cottony cushion scale in the U.S.A. It is a colourless, volatile liquid boiling point 26°C. and melting point about 14°C with a bitter almond odour. HCN is a gas at room temperature which has sp. gravity of 0.9483, hence is slightly lighter than air. Since the gas is lighter than air, the diffusion is mainly upward and outward. Under normal atmospheric pressure the gas is not able to penetrate closely packed materials and so partial vacuum fumigation is employed. There are four principal methods of applying this gas in any enclosure.

By the Pot Method: In this method the gas is generated from earthen jars or barrels placed within the enclosure to be fumigated by mixing water and sulphuric acid in the container. When the mixture is still hot, as a result of themical action, the proper amount of sodium or potassium cyanide is added which produces HCN gas. This method is not in much use.

$NaCN + H_2SO_4$--------------$NaHSO_4$ + HCN

The mixture may be used in the following ratio:

(a) For houses and godowns:

Water: 2 lit.

Sulphuric acid: 1 lit.per 30 cu m. space

Sodium cyanide: 1/2 kg

(b) For plants:

Water: 12 ml.

Sulphuric acid: 6 ml.per 30 cu. m.

Sodium cyanide: 4 g

Machine Method: Here the measured quantity of H_2SO_4 is introduced to a solution of Na CN in a special chamber or machine. In this way successive charges of gas are generated and lid from the machine into the enclosure to be fumigated through pipes leading from the machine.

Liquid Method: In this method, HCN is stored in liquid form in steel containers and is pumped out in the desired quantity through pipes and nozzles to the enclosure to be treated. 13 g liquid HCN is equivalent to 28 g of NaCN and is used @ 20 mg/30 cu.m. of space. This method is generally used in cases where large amount of gas is needed at one time.

The Dry Method: In this method $Ca(CN)_2$ or liquid HCN absorbed in inert earths or crude paper discs are sold in hermetically sealed cans. HCN is generated by spreading the material over the floor or blowing it into the atmosphere, when it readily gives up free hydrogen cyanide.

$Ca(CN)_2$ + $2H_2OCa(OH)2$ + 2HCN

$CaH_2(CN)_4$ + $2H_2OCa(OH)$ + 4HCN

Generally two principal types of calcium cyanide are available in the market, one is commonly used in the form of Cyano gas which contains 40-50% CaCN and gives its gas very slowly. Another form of dust is Cal cyanide' which contains 88% CaCN and gives of its gas very rapidly upon exposed to the moist air.

HCN is one of the most deadly gases, therefore, should never be undertaken by inexperienced persons. Proper gas masks should be used.

Carbon Bisulphide (CS_2)

It is also known as carbon disulphide and first employed in France to control pests of stored grains. It is a colourless, bad smelling liquid (due to traces of H_2S), slightly soluble in water (0.22 g/100 g at 22°C) and soluble in organic solvents. The gas has a sp. gr. of 2.63 and is highly inflammable and explosive in mixture above 1% in air. The gas is about two and a half times heavier than air, therefore, has considerable penetrative power. It is used for the control of stored product pests in godowns, and pests of cloths etc. Its emulsion is used for controlling soil insects and nematodes. Best result from fumigation with CS_2 to each 100 cub. ft of space or 7-12 lbs/ 370 qtls of grains. It should not be used directly on seeds, but it should be first poured on gunny sacks, rags etc., the latter being kept on the heap of grain. The gas flows down on account of its being heavier than air and penetrates into the heap. It is toxic to green plants, therefore, should not be used for fumigation of nursery material.

To protect against fire hazards, commercial mixture of 20% CS_2 with 80% carbon tetrachloride are available. One such mixture is sold under the trade name of Grain-ocide which is used @ 1.5 to 2 lb/1000 cu. ft of space or 12-20 lbs/370 qtls of grain.

Carbon Tetra Chloride (CCl_4)

This is thin, colourless, oily compound with 76°C boiling point. The gas has a pungent odour and is not inflammable or explosive. It is almost insoluble in water but is soluble in organic solvents. It is slow acting fumigant of comparatively low toxicity to insect and hence is not used alone but mixed with other much more toxic gases such as ethylene dichloride, carbon bisulphide to make them less explosive and reduce fire hazard. It increases the volatility and distribution of fumigants like methyl bromide, ethylene dichloride and chloropicrin.

Chloropicrin (CCl_3NO_2)

It is also known as trichloronitromethane and is one of the year gases possessing powerful irritating effect on the eyes. At room temperature it is colourless to yellowish liquid, boiling point 112.1°C. The sp. gravity of the gas is 5.7 and, therefore, has remarkable penetrative power. It is not inflammable, corrodes metals slightly but does not bleach or tarnish. It has been chiefly used in flour mills, grain elevators and farmers bins and against soil insects nematodes etc. It should not be used on growing plants or germinating seeds. Chloropicrin is mixed @. 2% with methyl bromide to act as indicator. It is used @ 1 lit./30 cu. m of space.

Ethylene Dibromide ($C_2H_4Br_2$)

It is 1, 2-dibromide ethane and is a colourless liquid with boiling point 131.6°C. The gas has a sp. gravity of 6.5 and slightly soluble in water but soluble in most organic solvents. EDB is odourless non-corrosive and non-inflammable liquid available as 5-20% solution. As a soil fumigant it is used for the control of wire worms, white grubs and nematodes @ 18 to 72 kg/hect. It is toxic to many plants, therefore, the treated soil should thoroughly be aerated before planting. It is highly toxic to warm blooded animals and can be absorbed directly through the skin. It is formulated as liquid, granule and even as capsule.

Ethylene Dichloride (CH_2ClCH_2Cl)

It is 1, 2-dichlroethane and is colourless liquid, very slightly soluble in water with boiling point of 83.7°C. The sp. gravity of gas is 3.4 and is inflammable between 6% and 16% in air. It is slow in its action and death of insects take place from one to three days after exposure. Since it is highly inflammable, therefore, used in mixture after the addition of about one-third (3: 1) of its volume of carbontetrachloride. The mixture is entirely free from fire hazard either as liquid or as gas. In India it is sold under the trade name of killoptera. The mixture is directly poured upon the material to be fumigated and allowed to vapourise. The mixture could be transported or stored in cans and easy to handle in the case of grain stores. The dosage for fumigation of grains etc. is 4 to 5 kg/100 quintets with an exposure period of 48 to 72 hours. The temperature during fumigation should be above 75°C. Since EDB is soluble in fats and oil, it is not recommended for use in commodities with high oil content. The mixture is injurious to the germination of most seeds at the recommended dosage.

Methyl Bromide (CH_3Br)

It is one of the most widely used fumigants because of its low boiling point (4.5°C). Methyl Bromide or Bromomethane is a colourless almost odourless gas at ordinary temperature. It is slightly soluble in water (1.34 g per 100 g at 25°C) but readily soluble in organic solvents and the gas has a sp. gravity of 3. 29. It requires a longer period of exposure as it kills the insect slowly. A dosage of 1.5 to 2 lb/1000 cu. ft. with an exposure period of 48 hours is advocated for stored grain pests. The gas is not inflammable and mixture with air can not be ignited by flame under conditions of ordinary fumigations. Since it is odourless, therefore, chloropicrin @ 2% is mixed with it to serve as an indicator. Bromomethane is useful for the fumigation of green houses, fruits and vegetables, stored food products, household goods and nursery stocks etc. The germination of seeds is not affected, if they are dry.

Naphthalene ($C_{10}H_8$)

It is white crystalline compound vapourises slowly to form a vapour which has a sp. gravity of 4.4. The gas has a pungent odour and is inflammable mixture with air. It is used against clothes moths and carpet beetles etc. Naphtalene is also used to a limited extent as a soil fumigant for wireworms and as a fumigant in green houses to control spider, mites and thrips. It is generally available as moth balls and moth flakes etc.

Parachloro Benzene ($C_6H_4Cl_2$)

It is a white crystalline material, vapourises slowly to form a inflammable gas. The gas has a sp. gr. of 5.1. Parachloro benzene is very slightly soluble in water and is soluble in organic solvents. It is used as a soil fumigant for controlling apple wooly aphis and larva of the peach tree borer and as a house hold fumigant and repellent for clothes moth and carpet beetle. It is also used to protect museum specimens from the attack of dermested beetle larvae.

Phosphine or Hydrogen Phosphide (PH_3)

It is a colourless, poisonous and inflammable gas which has been used as a fumigant for many years to fumigate grains, sacks and bins. The commercial product is available as phostoxin consists of 3 g tablets or pellets of ammonium carbamate mix the aluminium phosphide. Each 3 g tablet produces in the presence of moisture 1 g of phosphine by the following reaction:

$AlP + 2NH_4OCNH_2 + 3H_2O \rightarrow PH_3 + A1(OH)_3 + 2NH_3 — 2CO_2$

The CO_2 and NH_3 released by ammonium carbamate prevents spontaneous ignition of phosphine. The final product formed after decomposition is harmless Aluminium hydroxide. The gas is highly toxic to all stages of insect. Fumigation of packed bulk commodities can be carried and in an air-tight godown under gas proof sheet or in the container itself, if it is air tight. It does not affect the germination power of seeds. It is generally used @ I tablet per ton of produce with an exposure period of 5 days.

List of Insecticides registered on a regular basis under Section 9 (3) of the Insecticides Act, 1968 in India

Sl. No.	*Name of Insecticides*	*Sl. No.*	*Name of Insecticides*
1.	Acephate	2.	Alachlor
3.	Aldicarb	4.	Aliethrin
5.	Alphanapthyl acetic acid	6.	Aluminium phosphide
7.	Anilophos	8.	Arrazine
9.	Alphacypermethrin	10.	Aureofungin
11.	Barium carbonate	12.	Benzene hexachloride (BHC)
13.	Benthiocarb (Thiobencarb)	14.	Benomyl

Contd...

Sl. No.	*Name of Insecticides*	*Sl. No.*	*Name of Insecticides*
15.	Birertanol	16.	Bromadiolone
17.	Butachlor	18.	Calcium cyanide
19.	Captafol	20.	Captan
21.	Cartap hydrochloride	22.	Carbaryl
23.	Carbendazim	24.	Carbofuran
25.	Carboxin	26.	Chlorbenzilate
27.	Chlorfenvinphos	28.	Chlormequat chloride (CCC)
29.	Chlorothalonil	30.	Chlorphyriphos
31.	Copper oxychloride	32.	Copper sulphate
33.	Coumachlor	34.	Coumatetralyl
35.	Cuprous oxide	36.	Cypermethrin
37.	Cyfluthrin	38.	Dalapon
39.	Decamethrin/Deltamethrin	40.	Dichlorodiphenyltrichlorethane (DDT)
41.	Dichloropropane and dichloro propenes mixture (DD mixture)	42.	Dichlorovos (DDVP)
43.	Dicofol	44.	Diflubenzuron
45.	Dimethoate	46.	Dinocap
47.	Dioxathion	48.	Diuron
49.	Dodine	50.	Diazinon
51.	Dieldrin	52.	2, 4-Dichlorophenoxy acetic acid (2,4-D, sodium amine and ester salts)
53.	Ethylene dibromide	54.	Edifenphos-carbon tetrachloride mixture (EDCT mixture 3: 1)
55.	Endosulfan	56.	Ethofenprox
57.	Ethepon/Etofenprox	58.	Ethylene dibromide (EDB)
59.	Ethyl mercury chloride	60.	Ethion
61.	Fenitrothion	62.	Fenarimol
63.	Fenobucarb (EPMC)	64.	Ferithion
65.	Fenvalerate	66.	Ferbam
67.	Fluchloralin	68.	Fluvalinate
69.	Formothion	70.	Fosetyl-AI

Contd...

Sl. No.	*Name of Insecticides*	*Sl. No.*	*Name of Insecticides*
71.	Gibberellic acid	72.	Glyphosate
73.	Iprodione	74.	Isoproturon
75.	Kitazin	76.	Lime sulphur
77.	Lindane (Gamma BHC)	78.	Malathion
79.	Maleic hydrazide (MH)	80.	Mancozeb
81.	Menazon	82.	Metaldehyde
83.	Methabenzthiazuron	84.	Methoxy ethyl mercury chloride (MEMC)
85.	Methyl bromide	86.	Methyl bromide + Ethylene dibromide (1: 1 and 3: 1)
87.	Methyl chlorophenoxy acetic acid (MCPA)	88.	Metalaxyl
89.	Metoxuron	90.	Methomyl
91.	Metribuzin	92.	Metolachlor
93.	Monocrotophos	94.	Mone sodium methane arsonate (M/SMA)
95.	Myclobutanil	96.	Methyl parathion
97.	Nickel chloride	98.	Nicothine sulphate
99.	Oxadiazon	100.	Oxyacarboxin
101.	Oxydemeton-methyl	102.	Oxyfluorfen
103.	Para dichlorobenzene	104.	Paris green (Copper acetoarsenite)
105.	Paraquat dichloride	106.	Pendimethalin
107.	Permethrin	108.	Phenthoate
109.	Phenyl mercury acetate (PMA)**	110.	Phorate
111.	Phosalone	112.	Phosphamidon
113.	Pirimiphos-methyl	114.	Profenofos
115.	Propanil	116.	Propetamphos
117.	Propoxur	118.	Pyrethrum
119.	Quinalphos	120.	Sevidol (4: Carbaryl and Gamma HCH)
121.	Simazine	122.	Sirmate
123.	Sodium cyanide	124.	Streptomycin
125.	Sulphur	126.	Temephos

Contd...

Sl. No.	*Name of Insecticides*	*Sl. No.*	*Name of Insecticides*
127.	Trichloro acetic acid (TCA)	128.	Triallate
129.	Trichlorphon	130.	Tricyclazole
131.	Tridemorph	132.	Trifluralin
133.	Thiometon	134.	Thiram
135.	Thiophanate methyl	136.	Triazophos
137.	Triadimefon	138.	Warfarin
139.	Zinc phosphide	140.	Zineb
141.	Ziram	142.	Hexaconazole
143.	Pretilachlor	144.	Propiconazole

List of Banned Pesticides

Sl. No.	*Name of Pesticides*	*Sl. No.*	*Name of Pesticides*
1.	Dibromochloropropane	2.	Endrin (DBCP)
3.	Pentachlornitrobenzene	4.	Pentachlorophenol (PCP) (PCNB)
5.	Texaphene	6.	Ethyl parathion
7.	Chlordane	8.	Heptachlor
9.	Aldlrin (*w. e.f.* 1.1.1994)	10.	Paraquat-di-methyl Sulphate
11.	Nitrofen	12.	Tetradifen
13.	Methomyl	14.	Phosphamidon

List of Pesticides not Approved for Use

SI. No.	*Name of Pesticides*	*SI. No.*	*Name of Pesticides*
1.	Calcium arsonate	2.	EPN
3.	Azinpho-mcthyl	4.	Mevinphos (Phosdrin)
5.	Lead arsenate	6.	2, 4, 5-T
7.	Carbophenothion	8.	Vamidothion
9.	Mephosfolen	10.	Azinphos ethyl

Contd...

Sl. No.	Name of Pesticides	Sl. No.	Name of Pesticides
11.	Binapacryl	12.	Dicrotophos
13.	Thiodemetera/Dosi	14.	Fention acetate
15.	Fention hydroxide	16.	Chinomethionate
17.	Ammonium sulphamate	18.	Leptaphos (Phosvel)

List of Pesticides Under Review by Government

Sl. No.	Name of Pesticides	Sl. No.	Name of Pesticides
1.	Alachlor	2.	Benomyl
3.	Copper acetoarsenite	4.	Diuron
5.	Ethyl mercury chloride	6.	Fenarimol
7.	Menazon	8.	Oxyflourfen
9.	Sodium methane arsonate	10.	Calcium cyanide
11.	Thiometon	12.	Triazophos
13.	Tridemorph	14.	Monocrotophos
15.	Ziram	16.	Zineb

Newer Approaches in Insect Control

Pheromones: The term pheromone is generally used to those chemicals which are secreted into the external environment by an animal and that elicit a specific reaction in a receiving individual of the same species. These chemicals have also been known as ectohormones. Pheromones are divided into two groups:

(i) Which gives releaser effect; this impose a more or less immediate and reversible change in the behaviour of the recipient ?

(ii) Which gives as prima effect; this triggers a chain of physiological events in the receiving organisms ?

The latter class, which generally operates through gustatory stimulation by the chemicals and control such phenomenon as caste determination and reproductive control in social Hymenoptera (Ants and Bees) and Isoptera (Termites).

Mode of Action

Releasing pheromones are typically odorous and act directly on the central nervous system of the recipient. General types of behaviour that may be released include trial following, alarm, sexual activity, aggregation, dispersal and territoriality. From insect point of view two classes of pheromones that promote aggregation will be important.

Sex Pheromones: Which are released by one sex only and trigger behaviour patterns in the other sex that facilitate mating etc. These are also known as sex attractant or sex lures, *e.g.*, gyptol.

General Aggregation Pheromones: Which may be released by only one sex but cause approach responses by individuals of both sexes of the species.

Uses of Pheromones in Insect Control

In insect control only the sex pheromones are used. Sex elicited by insects have been isolated and identified chemically and they are being artificially synthesised in the laboratories of many advanced countries like U.K., U.S.A., etc. Sex pheromones thus prepared are available in the form of pheromone dispensers of several types. They are species specific and separate dispensers are required for separate insects. These dispensers are fitted into the pheromone traps which are of different types. The sex pheromones of female insects are available in the synthetic forms of pheromonal materials (dispensers) which attract the males of same species for mating. The purpose and use of pheromone traps are summarised below:

Monitoring of Insect Pests: The most important use of pheromone traps is for monitoring purposes. The outbreak of particular insect species may be ascertained at very early stage of its infestation and thus their population may be controlled at very appropriate time. In case of gram pod borer, Heliothis armigera, 4-6 moths per trap per night, warrants the suitable control measures. In this type of trapping (monitoring) four to six pheromone traps/hectare are recommended.

Mass Trapping and Destruction: Male moths may be captured in large numbers with the help of pheromone traps and thus their population explosion be brought down to considerable level, economically. Mass trapping should be done on campaign basis. Usually, 6-9 traps are used in one hectare for such purpose.

Mating Disruption Technique or Male Confusion Technique: In this technique the high concentrations of pheromonal material is used so that percipient males become confused and there becomes mating disruption between males and females. Sometimes the antennal (the organ of pheromone perception) loose the power of pheromone perception due to some high concentration of pheromonal material. Thus the male copulates with some unwanted plant materials. In this technique some micro encapsulated material is also used which is dispersed with cultivated field through the helicopter or aeroplanes. Thus there is inhibition of mating and so the multiplication of insect pests is checked to a some extent.

Insect Survey and Collection: Pheromone traps are also instrumental in the survey and collection of a particular insect species in the targeted area. It does not require the technical skill and expertise for the identification and collection of desired species.

Lure: Is the mechanism through which insects are directed through food mating/oviposition sites and aggregating. It operates through specific chemical (S) entitles, *e.g.*, methyl euginol, floral scent, carbon dioxide, lactic acid, sugar and propionitrol.

Hormones: In recent years hormones are also successfully used in minimising the pest population. The prime candidate for developing hormonal pesticides is the juvenile hormone that all insects secrete at certain stages in their lives. This hormone is secreted by corpora allata which is in the form of two tiny glands in the head. Ecdysone is another hormone, secreted from thoracic gland which is responsible for pupation and maturation in insects. These hormones have been shown to change the course of development in insects abruptly when applied at appropriate time and in turn these may be used as pesticides. Juvenile hormone increases the larval instrar when applied which may never pupate; besides, the chemosterilization property of this harmone has also been reported. By the application of Ecdyson hormone insect could be killed or may be brought in developing conditions during wrong season thus committing ecological suicide.

Antifeedant: The idea that a chemical could prevent feeding is not new, early workers usually referred to such materials as repellents. Antifeedants are not repellent in the true sense of the word, since insects are not driven away or kept away. Such chemicals inhibit feeding of the insects and thus protect the crop or commodity and eventuality starve to death without eating the leaves. There are following types of chemicals which show antifeeding property.

Triazene — Compound 24055

Organotins — Breston, Brestenol

Carbamate — Baygon

Botaniical extracts — Pyrethrins, Neem seed etc.

Mode of Action

Insect feeding consists of several stages these are:

(i) Orientation of attraction

(ii) Bitting.

(iii) Swallowing or Sustained feeding.

When insects are given the choice between antifeedant treated or untreated plants, there is little or no difference in initial attractiveness and equal numbers will be found on both groups of plants. Bitting takes place of both groups, but the differents lies in whether, there is a continuous bitting or a sustained feeding response. Those on treated plants continue taking tiny bites from the leaf as they wander around, where as those on untreated plants will feed in one spot unless disturbed.

In order to have sustained feeding it requires:

(i) The presence of gustatory stimulus.

(ii) The absence of any inhibitory stimulus.

(iii) and it must be hungry.

Evidence indicates the 1st of these are the point at which an antifeedant acts. Apparently antifeedant inhibit the taste receptors of the mouth region so that, lacking the proper gustatory stimulus, the insect fails to recognise the treated leaf as food and thus continuous foraging.

Chemosterilants: Chemicals which produce sterility and not killing the pest, are referred as chemosterilants. The basic concept behind the use of chemosterilants is to minimise the reproductive potential of the insects. There are two principles involved in insect control.

(i) Mass production, sterilization and release of sterile organism so that they will interfere with the remaining natural fertile population.

(ii) Sterilization of the required portion of the natural population by appropriate means (baiting etc.) so that the sterilize insects can not reproduce and at the same time they compete with the remaining fertile individuals and check the reproduction.

The sterilizing action of a chemical may be due to the fact that:

(i) It damages the chromosomes and leads to the formation of dominant lethal mutations in the semen of males or mature eggs of females;

(ii) It causes perishing of the cells in the premeiotic stages or aspermia;

(iii) It inactivates the sperm;

(iv) Under the influence of the toxicant, females lose the ability of laying eggs.

The practical application of chemosterilants depends on the following three points ;

(i) Nature of the active compounds.

(ii) Mode of administration.

(iii) Diversity of insect.

Types of Chemosterilants

Chemosterilants may be divided into three groups:

Alkylating agents: Specially effective as the male sterilants.

(a) Aziridines and Non-aziridines, *e.g.*, Tepa, Metepa, thiotepa, apholate.

(b) Sulfuric acid esters.

(c) Nitrogen mustards.

Antimetabolites: Mostly female sterilants, e.g.,

(a) Analogues of purines, pyrimictines.

(b) Folic acid.

(c) Analogues of amino acids.

Miscellaneous: Compounds

Examples: Hempa, colchicine, caomarin.

Mode of administration: Chemosterilants may be applied in the following manner:

(i) Direct contact with sprays and dusts.

(ii) Combination of insecticide and chemosterilant.

(iii) Oral ingestion.

(iv) Sterile male release method.

Advantages

The application of chemosterilants has following advantages:

Bonus effect: If the sexual competitiveness of an insect is maintained the fertile organisms will not be incapable of reproducing but have the ability to reproduce the reproductive capacity of the normal unsterilized population.

Space Effect: The sterile individuals through movement can effect the reproductive potential of the individuals outside the treated area.

Time Effect: The insects which are long lived and sterilized, they can effect the reproductive potential of the population in subsequent generation by mating competition.

Population Crush: The sterilized individuals will compete with unsterilized individuals for food and shelter in the environment during their normal surviving period. They also mate with even the hiding individuals.

6

Respiratory Qualities in Plants

Process of Respiration

Perhaps the term respiration was used with reference to the plants for the first time by J. Van Sachs in 1887. At that time respiration (rather called normal respiration) meant a particular kind of aerobic catabolism in which carbohydrates were completely oxidized by the protoplasm of plants and animals, while O_2 was used and CO_2 was liberated in equimolar amounts. Later on, it was realised that not only carbohydrates but rather stored food such as fats, proteins and organic acids also could undergo oxidative catabolism and that the ratio of CO_2 evolved to O_2 consumed was not always unity. By the end of the nineteenth century, experimental studies normally in the saprophytic micro-organisms, by W. Pfeffer, F. Czapeck, V.I. Palladin and many others, revealed that a variety of organic substrates underwent oxidative degradation to provide energy for biosynthetic reactions.

The discovery and isolation of the enzyme zymase from yeast by Buchner brothers, lead to the belief that oxidative fermentation could occur outside the living cells as well. Many other plant physiologists engaged in the study of intermediary metabolism of respiration and fermentation, assisted in the elucidation of metabolic pathway of respiration. The EMP pathway of glycolysis, Szent-Gyorgyi's dicarboxylic acid cycle and then Kreb's tricarboxylic acid cycle, which were discovered primarily in the animal tissues, were found to operate in plant tissues as well.

Respiration may be defined as the oxidative degradation of organic compounds to yield useful form of energy. The energy liberated is usually stored in the form of high energy molecules, adenosine triphosphate (ATP), which are regarded as energy currency of any living cell. These molecules are used up for driving various energy requiring reactions. Thus, respiration is the key physiological process in any living system, providing required energy for other processes.

In higher plants, the process of respiration involves the consumption of oxygen; the electrons coming out of the organic compounds move through other organic molecules to oxygen ultimately. The most common compound to be oxidized in the plants is glucose. One molecule of glucose yields 6 moles of CO_2 and 6 moles of H_2O and about 36 ATP molecules are synthesized from the energy liberated in the process;

$$C_6H_{12}O_6+6O_2 \rightarrow 6CO_2+6H_2O+\text{Energy (29kJ)}$$

$$\downarrow$$

$$36 \text{ ATP}$$

If we look at the overall chemical reaction, the process of respiration appears to involve both oxidation as well as reduction, which is a kind of rule of thumb in any chemical process. In this process, the organic substrates are oxidized to CO_2, and O_2 is reduced to water. Whatever, the mechanics of chemistry may be, the objective of the respiration is to liberate usable energy which is must for organism's life.

As the current status goes, the carbon reactions of respiration involve three distinct processes;

(1) *Glycolysis.* (Derived from the Greek words *glycos*, sugar (sweet) and *lysis*, dissolution) also called Embden-

Meyerhof-Parnas (EMP) pathway, named after the three principal scientists. This reaction is common to both fermentation (or anaerobic respiration) and aerobic respiration.

In the pathway, a hexose molecule such as glucose or fructose, is anaerobically converted to two molecules of pyruvic acid.

(2) *Decarboxylation*, converting pyruvic acid to acetyl Co-enzyme A (CoA).

(3) *The tricarboxylic acid cycle* (TCA cycle), or citric acid cycle, also called as Kreb's cycle for the famous British physiologist, Hans A. Krebs, who along with W.A. Johnson first described the cycle in 1937. In this cycle, acetyl CoA is oxidized to produce CO_2 and water. NADH and $FADH_2$ are also produced in the reaction.

An important pathway of hexose catabolism that bypasses the EMP pathway is *hexose monophosphate shunt* or *pentose shunt*.

Glycolysis

Glycolysis is the initial event in any kind of respiratory metabolism. It is the sequence of reactions that converts glucose into pyruvate with the concomitant production of ATP. The storage carbohydrates such as sucrose and starch, which serve as normal substrates for respiration are hydrolysed by the appropriate enzymes to their mono-saccharide units, glucose and fructose. The reactions of glycolysis start with glucose or fructose and operate throughout in the cytosol. In the first step, glucose is phosphorylated to produce glucose 6- phosphate by the enzyme hexokinase. One molecule of ATP is utilized in this process;

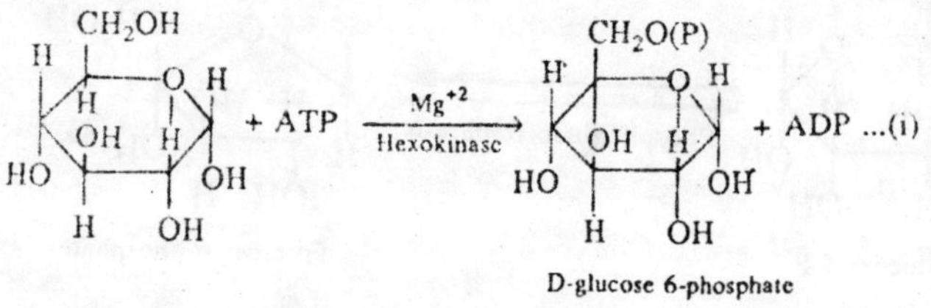

$$\text{Glucose } (CH_2OH) + ATP \xrightarrow[\text{Hexokinase}]{Mg^{+2}} \text{D-glucose 6-phosphate } (CH_2O(P)) + ADP \quad ...(i)$$

D-glucose 6-phosphate

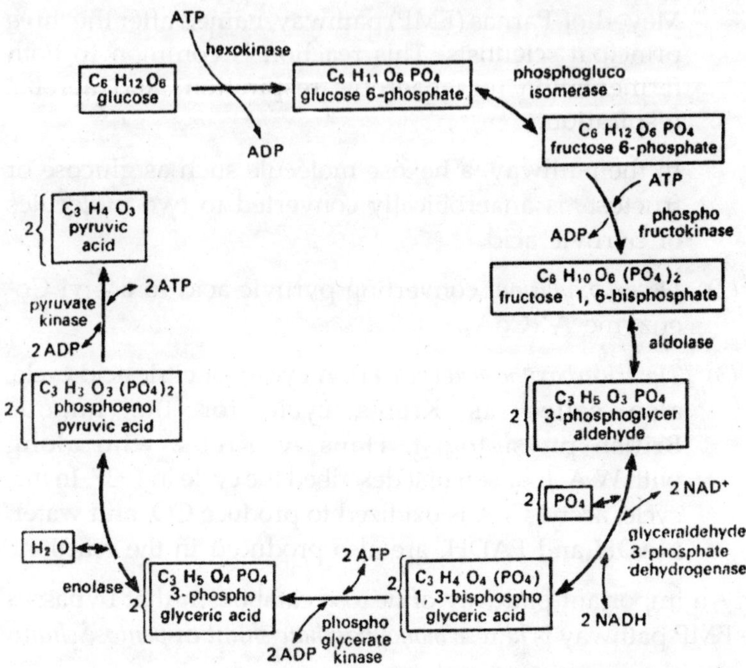

The metabolic pathway of glycolysis. [Numerals before each molecule indicate the number of molecules in the reaction]

Glucose 6-phosphate is then converted to its ketoisomer fructose 6-phosphate by the enzyme phosphogluco-isomerase (or phosphohexo isomerase). Although, the reaction is reversible, the equilibrium favours the synthesis of fructose 6-phosphate, as at a given time the amount of glucose 6-phosphate is about twice to that of fructose 6-phosphate. When fructose is the respiratory substrate, it can enter the glycolytic pathway at this step, as its phosphate;

$$\text{Glucose 6-P} \underset{}{\overset{\text{Phosphoglucoisomerase}}{\rightleftharpoons}} \text{Fructose 6-phosphate} \quad \text{...(ii)}$$

In the subsequent reaction, fructose 6-phosphate is phosphorylated to form bisphosphate derivative of fructose, by the enzyme phosphofructo-kinase;

$$\text{Fructose 6-P} \xrightarrow[\text{Phosphofructo kinase}]{Mg^{+2}} \text{Fructose 1,6-bisphosphate} \quad ...(iii)$$

The enzyme phosphofructokinase seems to be the rate limiting enzyme in the glycolytic pathway. It is inhibited by ATP, phosphoenol pyruvic acid and citric acid. The inhibition is relieved by inorganic phosphate and fructose 6-phosphate. Thus, when the potential end products of glycolysis and respiration such as phosphoenol pyruvic acid, citric acid and ATP are built up, this enzyme is subjected to the end product inhibition. In the next step, fructose 1,6-bisphosphate is cleaved by the enzyme i aldolase into two triose phosphates', a molecule of glyceraldehyde 3-phosphate. (GAP) and a dihydroxy acetone phosphate molecule. Dihydroxy acetone phosphate and GAP are in equilibrium through the enzyme triose phosphate isomerase;

$$\text{Fructose 1,6-bis-P} \underset{\text{Aldolase}}{\rightleftharpoons} \underset{\text{Dihydroxy acetone phosphate}}{CH_2O(P){-}C{=}O{-}CH_2OH} + \underset{\text{GAP}}{CHO{-}CHOH{-}CH_2O(P)} \quad ...(iv)$$

$$\text{Dihydroxyacetone phosphate} \underset{\text{Triose phosphate isomerase}}{\rightleftharpoons} \text{Glyceraldehyde 3-Phosphate} \quad ...(v)$$

Triose Phosphate Isomerase

In the subsequent reaction, the enzyme glyceraldehyde 3-phosphate dehydrogenase catalyses the oxidation of 3-phosphoglyceraldehyde to 1,3 bisphosphoglyceric acid. This reaction requires inorganic phosphate (Pi) and reduces NAD^+

to NADH. The two GAP molecules require two Pi molecules and two NADH molecules are produced;

$$\underset{\text{Glyceraldehyde 3-P}}{\begin{array}{l}CHO\\ |\\ CHOH\\ |\\ CH_2O(P)\end{array}} + NAD^+ + Pi \xrightleftharpoons[]{\text{Glyceraldehyde 3-phosphate dehydrogenase}} \underset{\text{1,3-bisphosphoglyceric acid}}{\begin{array}{l}COO(P)\\ |\\ CHOH\\ |\\ CH_2O(P)\end{array}} + NADH + H^+ \quad ...(vi)$$

Subsequently, one phosphate group of 1,3-bisphosphoglyceric acid is transferred to ADP, converting it to ATP; and producing 3-phosphoglyceric acid.

The production of ATP from ADP in this reaction (without involving oxidation of NADH) is called *substrate level phosphorylation.* The reaction is catalysed by the enzyme phosphoglycerate kinase. The 3-phosphoglyceric acid is mutated to 2-phosphoglyceric acid by the enzyme phosphoglycero-mutase;

$$\underset{\text{1,3-bisphosphoglyceric acid}}{\begin{array}{l}COO(P)\\ |\\ CHOH\\ |\\ CH_2O(P)\end{array}} + ADP \xrightarrow[\text{Phosphoglycerate kinase}]{Mg^{+2}} \underset{\text{3-phosphoglyceric acid}}{\begin{array}{l}COOH\\ |\\ CHOH\\ |\\ CH_2O(P)\end{array}} + ATP \quad ...(vii)$$

$$\underset{\text{3-phosphoglyceric acid}}{\begin{array}{l}COOH\\ |\\ CHOH\\ |\\ CH_2O(P)\end{array}} \xrightleftharpoons[\text{Phosphoglyceromutase}]{} \underset{\text{2-phosphoglyceric acid}}{\begin{array}{l}COOH\\ |\\ CHO(P)\\ |\\ CH_2OH\end{array}} \quad ...(viii)$$

In the next step, 2-phosphoglyceric acid is dehydrated to produce phosphoenol pyruvic acid by a Mg^{+2} dependent enzyme enolase;

$$\underset{\text{2-phosphoglyceric acid}}{\begin{array}{l}COOH\\ |\\ CHO(P)\\ |\\ CH_2OH\end{array}} \xrightarrow[\text{Enolase}]{Mg^{+2}\ H_2O} \underset{\text{Phosphoenol-pyruvic acid}}{\begin{array}{l}COOH\\ |\\ CO(P)\\ \|\\ CH_2\end{array}} \quad ...(ix)$$

Phosphoenol pyruvic acid can be hydrolysed yielding high energy and thus phosphorylating ADP to ATP. This is another step of substrate level phosphorylation in glycolysis. The reaction is catalysed by Mg^{+2} dependent pyruvate kinase;

$$\underset{\text{Phosphoenol-pyruvic acid}}{\begin{array}{l}COOH\\ |\\ CO(P)\\ \|\\ CH_2\end{array}} + ADP \xrightarrow[\text{Pyruvate kinase}]{Mg^{+2}} \underset{\text{Pyruvic acid}}{\begin{array}{l}COOH\\ |\\ C{=}O\\ |\\ CH_3\end{array}} + ATP \qquad ...(x)$$

The overall reaction for glycolysis (steps i to x) can be written as follows:

Glucose + 2ADP + 2Pi + $2NAD^+$ $\rightarrow$ 2 Pyruvic acid + 2ATP + 2NADH + $2H^+$ + $2H_2O$

There is a net production of 2 ATP and two NADH molecules (Table). The NADH can be oxidized in the mitochondria to produce further ATP molecules by oxidative phosphorylation.

Consumption and Generation of ATP in Glycolysis

Reaction	*ATP change per glucose*
Glucose $\rightarrow$ Glucose 6-phosphate	- 1
Fructose 6-phosphate $\rightarrow$ Fructose 1,6-bisphosphate	- 1
2 [1,3-bisphosphoglycerate] $\rightarrow$ 2 [3-phosphoglycerate]	+ 2
2 [Phosphoenol pyruvate] $\rightarrow$ 2 [pyruvate]	+ 2
Net + 2	

A summary of reactions of glycolysis is given in table.

Reactions of Glycolysis.

Step	Reaction	Enzyme	Type of reaction
1. Glucose + ATP	Hexokinase glucose 6-phosphate + ADP + H^+	Phosphoryl	transfer
2. Glucose 6-phosphate	Phospho-glucose — fructose 6-phosphate	Isomerization isomerase	
3. Fructose 6-phosphate + ATP	Phospho → fructose 1,6-bisphosphate + ADP + H^+	Phosphoryl fructokinase	transfer
4. Fructose 1,6-bisphosphate	Aldolase — dihydroxy acetone phosphate + glyceraldehyde 3-phosphate	Aldol cleavage	
5. Dihydroxy acetone phosphate	Triose phosphate — glyceraldehyde 3-phosphate	Isomerization isomerase	
6. Glyceraldehyde 3-phosphate + NAD^+ + Pi	Glyceraldehyde — 1,3-bisphosphoglycerate + NADH + H^+	Phosphorylation 3-phosphate dehydrogenase	coupled to oxidation
7. 1,3-bisphosphoglycerate + ADP	Phosphoglycerate — 3-phosphoglycerate + ATP	Phosphoryl kinase	transfer
8. 3-phosphoglycerate — 2-phosphoglycerate	Phosphoglycero-	Phosphoryl shift mutase	
9. 2-phosphoglycerate	Enolase — phosphoenol pyruvate + H2O	Dehydration	
10. Phosphoenol pyruvate + ADP + H^+	Pyruvate kinase → pyruvate + ATP	Phosphoryl	transfer

Decarboxylation of Pyruvate

Decarboxylation of pyruvic acid to produce a two carbon compound the acetate, occurs in the mitochondrial matrix and it may be considered as the initial reaction in aerobic respiration. The reaction is catalysed by an enzyme system called pyruvate dehydrogenase complex which is in fact composed of three enzyme units: (i) Pyruvate decarboxylase, (ii) Lipoate acetyl transferase and (iii) Lipoate dehydrogenase. The entire sequence requires five cofactors, Mg^{+2}, Coenzyme A (CoA), lipoic acid and thiamine pyrophosphate (TPP).

In the first step, the enzyme pyruvate decarboxylase forms TPP derivative of acetate, and CO_2 is released. The reaction requires Mg^{+2} and

COOH | C=O | CH_3 (Pyruvic acid) + TPP —(Pyruvate decarboxylase)→ (TPP) ‖ C—OH | CH_3 (TPP-acetate) + CO_2

In the next step, TPP is displaced by the lipoic acid and free TPP is liberated. The reaction is catalysed by lipoate acetyl transferase;

(TPP) ‖ C—OH | CH_3 (TPP-acetate) + Lipoic acid —(Lipoate acetyl transferase)→ (Lip) | S | C=O | CH_3 (Acetyl hydrolipoate) + TPP

In the last step, the acetate from acetyl dihydrolipoate is displaced by coenzyme A to yield reduced lipoic acid and acetate. The enzyme for this reaction is lipoate dehydrogenase;

(Lip) | S | C=O | CH_3 (Acetyl hydrolipoate) + CoASH —(Lipoate dehydrogenase)→ CH_3CO—S—CoA (Acetyl Coenzyme-A) + Reduced lipoic acid

However, reduced lipoic acid is not the final product of pyruvate decarboxylation. Another associated enzyme called dihydrolipoate dehydrogenase catalyses the transfer of electrons from reduced lipoic acid to NAD^+ through FAD. In fact the enzyme dihydrolipoate dehydrogenase is a flavoprotein with tightly bound FAD. The FAD is first reduced to $FADH_2$ and then $FADH_2$ reduces NAD^+ to NADH;

$$\text{Reduced lipoic acid} + \text{FAD} \xrightarrow{} \text{Lipoic acid} + \text{FADH}_2$$

$$\text{FADH}_2 + \text{NAD}^+ \xrightarrow[\text{dehydrogenase}]{\text{Dihydrolipoate}} \text{FAD} + \text{NADH} + \text{H}^+$$

Thus, the decarboxylation of pyruvic acid to acetyl coenzyme A produces NADH. Two molecules of NADH will be produced from one molecule of a hexose. This can be oxidized to yield ATP by oxidative phosphorylation. The overall reaction of decarboxylation of pyruvic acid can be written as follows :

$$\text{Pyruvic acid} + \text{CoASH} + \text{NAD}^+ \xrightarrow[\text{dehydrogenase complex}]{\text{Pyruvate}} \text{Acetyl CoA} + \text{NADH} + \text{CO}_2$$

Tricarboxylic Acid Cycle

The acetyl coenzyme A is oxidized to CO_2 and water, yielding NADH and $FADH_2$ through a cyclic process *i.e.*, one of the products (oxaloacetic acid) is channelled back to the process for generating more CO_2 and NADH/$FADH_2$. In the late 1940's A.L. Lehninger demonstrated that isolated mitochondria could oxidise pyruvic acid to CO_2 and H_2O with the consumption of O_2. Mitochondria were then shown to contain all the enzymes of tricarboxylic acid cycle and the pyruvate dehydrogenase complex. It is therefore logical to conclude that *in vivo* oxidation of pyruvic acid, through TCA cycle takes place in the mitochondria.

Reactions of tricarboxylic acid cycle are shown in figure a and b. In the first step, acetyl CoA condenses with one molecule of oxaloacetic acid to produce a tricarboxylic (containing 3 COOH group) acid, citric acid. Hence the name of the cycle is tricarboxylic acid cycle or citric acid cycle. The enzyme for this reaction is citrate synthetase;

$$\begin{array}{l} COOH \\ | \\ C{=}O \\ | \\ CH_2 \\ | \\ COOH \\ \text{Oxaloacetic} \\ \text{acid} \end{array} + \begin{array}{l} O \\ \| \\ C{-}SCoA + H_2O \\ | \\ CH_3 \\ \text{Acetyl CoA} \end{array} \xrightarrow[\text{synthetase}]{\text{Citrate}} \begin{array}{c} COOH \\ | \\ CH_2 \\ | \\ HOOC{-}C{-}OH + CoASH + H^+ \\ | \\ CH_2 \\ | \\ COOH \\ \text{Citric acid} \end{array}$$

Citric acid is then isomerised to isocitric acid by the enzyme aconitase. In fact in this isomerisation, an intermediate compound cis-aconitic acid is formed which remains tightly bound to the enzyme. Cis-aconitic acid is then hydrolysed to isocitric acid;

$$\begin{array}{c} COOH \\ | \\ CH_2 \\ | \\ HOOC{-}C{-}OH \\ | \\ CH_2 \\ | \\ COOH \\ \text{Citric acid} \end{array} \underset{\text{Acotinase}}{\overset{H_2O}{\rightleftharpoons}} \begin{array}{c} COOH \\ | \\ CH \\ \| \\ C{-}COOH \\ | \\ CH_2 \\ | \\ COOH \\ \text{cis-aconitic acid} \end{array} \underset{\text{Acotinase}}{\overset{H_2O}{\rightleftharpoons}} \begin{array}{c} COOH \\ | \\ CHOH \\ | \\ CH{-}COOH \\ | \\ CH_2 \\ | \\ COOH \\ \text{iso-citric acid} \end{array}$$

Next step is the first oxidative decarboxylation step of TCA cycle. In this reaction, isocitric acid is decarboxylated to yield 2-oxoglutaric acid in a two step reaction. The intermediate compound is oxalosuccinic acid. The reaction generates NADH also and the enzyme for the reaction is isocitrate dehydrogenase;

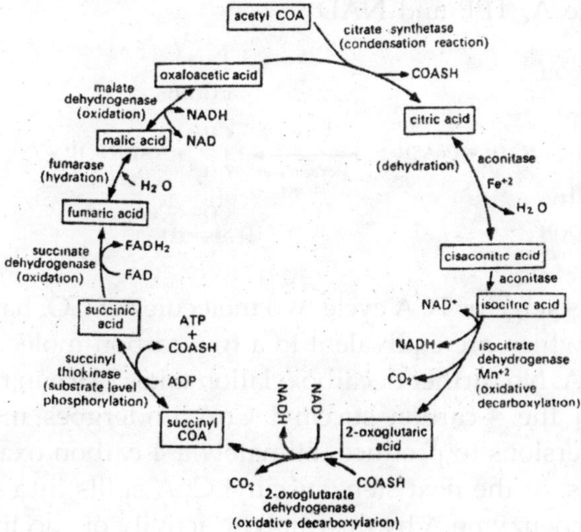

(a) The tricarboxylic acid (TCA) cycle

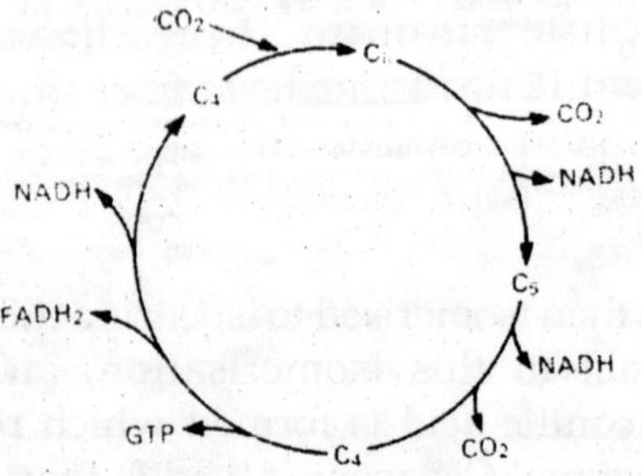

(b) An overview of the citric acid cycle.

$$\begin{array}{l} COOH \\ | \\ CHOH \\ | \\ CH{-}COOH \\ | \\ CH_2 \\ | \\ COOH \\ \text{iso-citric acid} \end{array} + NAD^+ \xrightarrow[\text{dehydrogenase}]{\text{Iso-citrate}} \begin{array}{l} COOH \\ | \\ C{=}O \\ | \\ CH_2 \\ | \\ CH_2 \\ | \\ COOH \\ \text{2-oxoglutaric acid} \end{array} + CO_2 + NADH + H^+$$

In the next step, 2-oxoglutaric acid is decarboxylated (second oxidative decarboxylation step) by an enzyme complex, 2-oxoglutarate dehydrogenase complex and produces succinyl CoA, CO_2 and NADH. Cofactors required for this reaction are Coenzyme A, TPP and NAD^+;

$$\begin{array}{l} COOH \\ | \\ C{=}O \\ | \\ CH_2 \\ | \\ CH_2 \\ | \\ COOH \\ \text{2-oxoglutaric acid} \end{array} + NAD^+ + CoASH \xrightarrow[\text{2-oxoglutarate dehydrogenase}]{\text{TPP.}} \begin{array}{l} COOH \\ | \\ CH_2 \\ | \\ CH_2 \\ | \\ CO{-}SCoA \\ \text{Succinyl CoA} \end{array} + NADH + H^+ + CO_2$$

At this stage of TCA cycle two molecules of CO_2 have been evolved, which are equivalent to a two carbon molecule, *e.g.*, acetyl CoA. No further decarboxylation takes place in the TCA cycle and the 4-carbon succinyl CoA undergoes metabolic interconversions to produce ultimately a 4-carbon oxaloacetic acid. Thus, in the next step, succinyl CoA splits into succinic acid and coenzyme A by the catalytic activity of succinyl CoA synthetase or succinate thiokinase. The reaction also generates ATP from ADP and Pi (substrate level phosphorylation). The enzyme can also phosphorylate GDP and IDP, but probably ADP is the normal substrate;

$$\underset{\text{Succinyl CoA}}{\begin{array}{l} COOH \\ | \\ CH_2 \\ | \\ CH_2 \\ | \\ CO{-}SCoA \end{array}} + ADP + Pi \xrightleftharpoons[\text{Succinyl CoA synthetase or Succinate thiokinase}]{Mg^{+2}} \underset{\text{Succinic acid}}{\begin{array}{l} COOH \\ | \\ CH_2 \\ | \\ CH_2 \\ | \\ COOH \end{array}} + ATP + CoASH$$

It is the only step in citric acid cycle that directly yields a high energy phosphate bond (substrate level phosphorylation).

Third oxidation in the cycle is unaccompanied with decarboxylation. In this reaction, succinic acid is oxidized to fumaric acid by the enzyme succinate dehydrogenase. The enzyme reduces FAD to $FADH_2$ simultaneously;

$$\underset{\text{Succinic acid}}{\begin{array}{l} COOH \\ | \\ CH_2 \\ | \\ CH_2 \\ | \\ COOH \end{array}} + FAD \xrightleftharpoons[\text{Succinate dehydrogenase}]{} \underset{\text{Fumaric acid}}{\begin{array}{l} COOH \\ | \\ CH \\ \| \\ HC \\ | \\ COOH \end{array}} + FADH_2$$

The enzyme fumarase, catalyses the hydration of fumaric acid, which is then converted to malic acid;

$$\underset{\text{Fumaric acid}}{\begin{array}{l} COOH \\ | \\ CH \\ \| \\ HC \\ | \\ COOH \end{array}} \xrightleftharpoons[\text{Fumarase}]{H_2O} \underset{\text{Malic acid}}{\begin{array}{l} COOH \\ | \\ CHOH \\ | \\ CH_2 \\ | \\ COOH \end{array}}$$

In the final reaction, (fourth oxidation), malic acid is oxidised to oxalacetic acid by the enzyme malate dehydrogenase. In this reaction, NAD^+ is reduced to NADH;

$$\underset{\text{Malic acid}}{\begin{array}{l} COOH \\ | \\ CHOH \\ | \\ CH_2 \\ | \\ COOH \end{array}} + NAD^+ \xrightleftharpoons[\text{Malate dehydrogenase}]{} \underset{\text{Oxaloacetic acid}}{\begin{array}{l} COOH \\ | \\ C{=}O \\ | \\ CH_2 \\ | \\ COOH \end{array}} + NADH + H^+$$

Production of oxaloacetic acid completes the cycle, as it can condense again with acetyl CoA to produce citric acid and so on. If we look at the overall reactions of TCA cycle, 3 molecules of NADH, one molecule of $FADH_2$, one molecule of ATP and two molecules of CO_2 are produced. Starting from acetyl CoA, the overall reactions of TCA cycle can be written as follows:

$$CH_3CO\ CoA + 3NAD^+ + FAD + ADP + Pi + 2H_2O$$

$$\rightarrow 2CO_2 + 3NADH + FADH_2 + ATP + 3H^+ + CoASH$$

Reduction of $FADH_2$ and NADH by O_2 in the electron transport system (next stage in aerobic respiration) produces water ultimately. In this process energy is also liberated which is used in the generation of ATP from ADP and Pi. The generation of ATP from ADP and Pi during oxidation of $NADH/FADH_2$ is called oxidative phosphorylation.

Pentose Phosphate Pathway: Until 1950, it was believed that glycolysis and TCA cycle were the only reactions by which plants could oxidise carbohydrates to obtain energy.

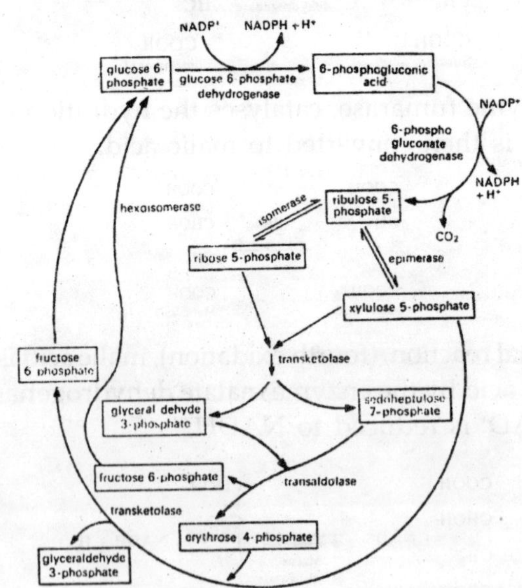

Pentose phosphate pathway

However, in early 1950's, studies conducted by M. Gibbs and by B. Axelrod and H. Beevers in USA established a different pathway of oxidative degradation of carbohydrates. In this pathway, the hexose sugars underwent catabolism through 5-carbon sugar intermediates and hence the pathway was called Pentose Phosphate Pathway (PPP). It has also been called the "Warburg-Dickens Pathway", the "Hexose Monophosphate Shunt" and the "Phosphogluconate Pathway". The reactions of the pathway take place in the cytosol and also in the chloroplasts.

In the first reaction, glucose 6-phosphate is oxidised to 6-phospho-gluconic acid by the enzyme glucose 6-phosphate dehydrogenase. $NADP^+$ is the electron acceptor in this reaction;

$$\begin{array}{l} CH_2O(P) \\ HOH \\ OHH \\ OHOH \\ HOH \\ \text{Glucose} \\ \text{6-phosphate} \end{array} + NADP^+ \xrightarrow[\text{dehydrogenase}]{\text{Glucose 6-phosphate}} \begin{array}{c} COOH \\ | \\ H—C—OH \\ | \\ HO—C—H \\ | \\ H—C—OH \\ | \\ H—C—OH \\ | \\ CH_2O(P) \\ \text{6-phospho} \\ \text{gluconic acid} \end{array} + NADPH + H^+ \quad \ldots(i)$$

The phosphogluconic acid does not accumulate; but undergoes oxidative (dehydrogenation) decarboxylation rapidly by 6-phosphogluconic acid dehydrogenase to yield a 5-carbon compound ribulose 5-phosphate. NADPH and CO_2 are also produced in the reaction;

$$\begin{array}{l} COOH \\ | \\ (CHOH)_4 \\ | \\ CH_2O(P) \\ \text{6-phosphogluconic} \\ \text{acid} \end{array} + NADP^+ \xrightarrow[\text{acid dehydrogenase}]{\text{6-phosphogluconic}} \begin{array}{c} CH_2OH \\ | \\ C{=}O \\ | \\ H—C—OH \\ | \\ H—C—OH \\ | \\ CH_2O(P) \\ \text{Ribulose} \\ \text{5-phosphate} \end{array} + NADPH + H^+ + CO_2 \quad \ldots(ii)$$

The two steps (i and ii) are irreversible reactions and they represent the only oxidative steps in PPP, and the CO_2 arising in step (ii) is the only CO_2 released in entire pathway. The subsequent reactions convert the pentose sugar (ribulose

5-phosphate) back to hexose sugar, involving some reactions which are common to Calvin cycle as well. Thus, in the next step, ribulose 5-phosphate is either isomerised by phosphoriboisomerase to ribose 5-phosphate or epimerised by phosphoribose epimerase to xylulose 5-phosphate;

$$\underset{\text{Ribulose 5-phosphate}}{\begin{array}{c} CH_2OH \\ | \\ C{=}O \\ | \\ H{-}C{-}OH \\ | \\ H{-}C{-}OH \\ | \\ CH_2O(P) \end{array}} \underset{\text{isomerase}}{\overset{\text{Phosphoribo}}{\rightleftharpoons}} \underset{\text{Ribose 5-phosphate}}{\begin{array}{c} CHO \\ | \\ (CHOH)_3 \\ | \\ CH_2O(P) \end{array}} \qquad \underset{\text{Phosphoribose epimerase}}{\rightleftharpoons} \underset{\text{Xylulose 5-phosphate}}{\begin{array}{c} CH_2OH \\ | \\ C{=}O \\ | \\ (CHOH)_2 \\ | \\ CH_2O(P) \end{array}} \quad \ldots(iii)$$

Subsequently, two ribose 5-phosphate and two xylulose 5-phosphate combine to form two molecules of sedoheptulose 7-phosphate. The enzyme catalysing this reaction is transketolase which requires Mg^{+2} and TPP as cofactros;

$$\underset{\text{Xylulose 5-phosphate}}{\begin{array}{c} CH_2OH \\ | \\ C{=}O \\ | \\ (CHOH)_2 \\ | \\ CH_2O(P) \end{array}} + \underset{\text{Ribose 5-phosphate}}{\begin{array}{c} CHO \\ | \\ (CHOH)_3 \\ | \\ CH_2O(P) \end{array}} \underset{\text{Transketolase}}{\overset{Mg^{+2},\ TPP}{\rightleftharpoons}} \underset{\text{Sedoheptulose 7-phosphate}}{\begin{array}{c} CH_2OH \\ | \\ C{=}O \\ | \\ (CHOH)_4 \\ | \\ CH_2O(P) \end{array}} + \underset{\text{Glyceraldehyde 3-phosphate}}{\begin{array}{c} CHO \\ | \\ CHOH \\ | \\ CH_2O(P) \end{array}} \quad \ldots(iv)$$

In the next step, two sedoheptulose 7-phosphate molecules react with the two glyceraldehyde 3-phosphate to form two molecules of erythrose 4-phosphate and two molecules of fructose 6-phosphate. The enzyme for this reaction is transaldolase;

$$\underset{\text{Sedoheptulose 7-phosphate}}{\begin{array}{c} CH_2OH \\ | \\ C{=}O \\ | \\ (CHOH)_4 \\ | \\ CH_2O(P) \end{array}} + \underset{\text{Glyceraldehyde 3-phosphate}}{\begin{array}{c} CHO \\ | \\ H{-}C{-}OH \\ | \\ CH_2O(P) \end{array}} \underset{\text{Transaldolase}}{\rightleftharpoons} \underset{\text{Fructose 6-phosphate}}{\begin{array}{c} CH_2OH \\ | \\ C{=}O \\ | \\ (CHOH)_3 \\ | \\ CH_2O(P) \end{array}} + \underset{\text{Erythrose 4-phosphate}}{\begin{array}{c} CHO \\ | \\ (CHOH)_2 \\ | \\ CH_2O(P) \end{array}} \quad \ldots(v)$$

In the next reaction, the enzyme transketolase again catalyses the reaction between two molecules of erythrose

4-phosphate and two molecules of xylulose 5-phosphate to form two molecules of glyceraldehyde 3-phosphate and two molecules of fructose 6-phosphate;

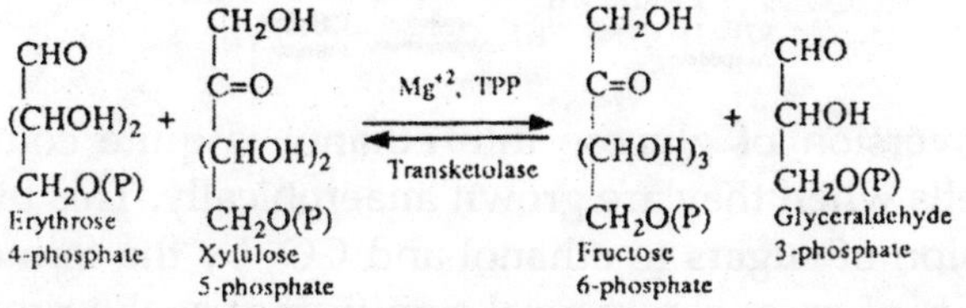

$$\begin{array}{c} CHO \\ | \\ (CHOH)_2 \\ | \\ CH_2O(P) \\ \text{Erythrose} \\ \text{4-phosphate} \end{array} + \begin{array}{c} CH_2OH \\ | \\ C{=}O \\ | \\ (CHOH)_2 \\ | \\ CH_2O(P) \\ \text{Xylulose} \\ \text{5-phosphate} \end{array} \underset{\text{Transketolase}}{\overset{Mg^{+2},\ TPP}{\rightleftharpoons}} \begin{array}{c} CH_2OH \\ | \\ C{=}O \\ | \\ (CHOH)_3 \\ | \\ CH_2O(P) \\ \text{Fructose} \\ \text{6-phosphate} \end{array} + \begin{array}{c} CHO \\ | \\ CHOH \\ | \\ CH_2O(P) \\ \text{Glyceraldehyde} \\ \text{3-phosphate} \end{array} \quad ...(vi)$$

Fructose 6-phosphate is then converted to glucose 6-phosphate by the enzyme phosphohexoisomerase. Glucose 6-phosphate can go to PPP again and the cycle is completed;

$$\text{Fructose 6-phosphate} \underset{\text{Phosphohesoisomerase}}{\rightleftharpoons} \text{Glucose 6-phosphate} \quad ...(vii)$$

[This is the same as reaction (ii) in glycolysis]

As is evident from the description, the pentose phosphate pathway is not as efficient as glycolysis/Krebs cycle in generating reducing power and energy. However, the pathway is important because it provides reduced NADPH for many biosynthetic reactions. Further, the metabolites ribose 5-phosphate and erythrose 4-phosphate produced in the pathway are important molecules as they are used in nucleic acid and lignin biosynthesis, respectively.

Anaerobic Respiration and Fermentation: As is evident for the earlier account in this chapter, the pyruvic acid produced in glycolysis enters the mitochondria, where it is oxidised by the TCA cycle. However, when O_2 supply is limiting, or in anaerobic organisms, pyruvic-acid is converted to either ethanol or lactic acid.

In the production of ethanol, pyruvic acid is decarboxylated to acetaldehyde through pyruvate decarboxylase enzyme, followed by acetaldehyde reduction to ethanol by the enzyme alcohol dehydrogenase;

$$\underset{\text{Pyruvic acid}}{\begin{matrix} CH_3 \\ | \\ C{=}O \\ | \\ COOH \end{matrix}} \xrightarrow[\text{decarboxylase}]{\text{Pyruvate}} \underset{\text{Acetaldehyde}}{\begin{matrix} CH_3 \\ | \\ CHO \end{matrix}} + CO_2$$

$$\underset{\text{Acetaldehyde}}{\begin{matrix} CH_3 \\ | \\ CHO \end{matrix}} + NADH + H^+ \xrightarrow[\text{dehydrogenase}]{\text{Alcohol}} \underset{\text{Ethanol}}{\begin{matrix} CH_3 \\ | \\ CH_2OH \end{matrix}} + NAD^+$$

Conversion of glucose into ethanol is quite common in yeast cells when they are grown anaerobically. This biological conversion of sugars to ethanol and CO_2 by the fermentation process is of great commercial significance as the process has been utilised to produce alcoholic drinks through the centuries. In the overall process, one molecule of glucose produces two molecules of ethanol, and two ATP Molecules are also generated;

Glucose + 2ADP + 2pi $\rightarrow$ 2 ethanol + $2CO_2$ + 2ATP + $2H_2O$

It should be realised that there is no net production of NADH in alcoholic fermentation. The two NADH Molecules generated during glycolytic conversion of glucose to pyruvic acid are consumed when pyruvic acid is reduced to ethanol. Fermantation also occurs in the cells of higher plants when O_2 is not enter the cells. In most water logged soils, root respiration is through anaerobic means or fermentation.

In many algae and fungi, pyruvic acid is converted to lactic acid instead of to ethanol under anaerobic conditions. Lactic acid is also produced in human specially in the muscle cells during strenuous physical exercises. However, the process is less common in higher plants. The enzyme catalysing lactic acid synthesis from pyruvic acid is lactate dehydrogenase. The creation uses NADH generated in glycolysis. So there is no net production of NADH in lactic acid fermentation, as well;

$$\underset{\text{Pyruvic acid}}{\begin{matrix} CH_3 \\ | \\ C{=}O \\ | \\ COOH \end{matrix}} + NADH + H^+ \underset{\text{dehydrogenase}}{\overset{\text{Lactate}}{\rightleftharpoons}} \underset{\text{Lactic acid}}{\begin{matrix} CH_3 \\ | \\ CHOH \\ | \\ COOH \end{matrix}} + NAD^+$$

Electron Transport System and Oxidative Phosphorylation : Although, the glycolytic and TCA cycle pathway produce some ATP by substrate level phosphorylation, most of the energy is conserved in the form of reductants NADH and $FADH_2$. These molecules are then oxidised by O_2 to produce water molecules. In the process, NADH (or $FADH_2$) donate electrons which are accepted by O_2,

$$NADH \rightarrow NAD^+ + e^- + H^+$$

$$2H^+ + 2e^- + \frac{1}{2} O_2 \rightarrow H_2O$$

However, neither NADH nor $FADH_2$ can combine directly to O_2, but the electrons from these molecules to O_2 are transported through a series of electron carriers. These electron carriers between NADH and O_2, constitute the electron transport system or cytochrome system of the mitochondria. They are all located in the inner mitochondrial membrane. These inner membranes are highly convoluted, protruding into the interior of the mitochondria into sheet like patterns in many places.

Each such convolution or folding is called *cristae.* They increase the total surface area where enzymes of electron transport system and TCA cycle are located and where these reactions take place. The mitochondria is self sufficient in carrying out all the reactions of TCA cycle and electron transport. This has been demonstrated by several scientists using isolated mitochondria from both plants and animals.

NADH is first oxidised to NAD^+ by a flavoprotein complex. The flavin (FMN) of the flavoprotein complex in turn is reduced to $FMNH_2$. However, both FMN and FAD can also exist as free radicals and accept electrons from NADH. The complex catalysing this reaction is often called complex I.

The reaction involves two electron transfer, and the energy generated in this transfer is large enough to synthesize one molecule of ATP from ADP and Pi. The electrons from flavoprotein (complex I) or from $FADH_2$ generated in the TCA cycle are transported to a specific quinone molecule; CoQ or ubiquinone. The CoQ is reduced and flavins are reoxidised.

The electrons from CoQ then move to cytochromes. The cytochromes are Fe containing proteins. The Fe ion undergoes valency changes (ferrous → ferric) during oxidation-reduction of cytochromes. There are three types of cytochromes in the mitochondria; a, b and c. They differ according to their absorption spectra. The electrons from CoQ are accepted by cytochrome b; cytochrome b than donates electrons to cytochrome c_1 (one of the two cytochrome c). The free energy change during this step of electron transfer (cytochrome b to c_1,) is large enough to generate a molecule of ATP. This step is inhibited by the antibiotic antimycin A. The electrons from cytochrome c_1 then move to cytochrome c and then to cytochrome $a_1 + a_3$ (a_1 and a_3 are two types of cytochrome a). There is a production of another molecule of ATP during electron transfer from cytochrome c to cytochrome $a_1 + a_3$. The cytochrome a complex ($a_1 + a_3$) is known as *terminal oxidase* or sometimes as *cytochrome oxidase*. This complex itself catalyses the transfer of electrons ultimately to O_2 which is reduced to form H_2O.

Thus, when both the electrons from NADH are transported to O_2, 3 ATP molecules are generated. However, when $FADH_2$ produced in glycolysis is oxidised only 2 ATP molecules are produced. Thus, in the presence of O_2 total number of ATP molecules produced is far more than what is produced in glycolysis and anaerobic respiration during O_2 deficiency.

Mechanism of Oxidative Phosphorylation: Generation of ATP from ADP and Pi in the mitochondria during electron transport from NADH to O_2 is known as oxidative phosphorylation. The mechanism of oxidative phosphorylation is same as that of photo-Phosphorylation except that it takes place inside the mitochondria instead of in chloroplast. The electron transport generates a pH gradient across the inner mitochondria membrane, as is generated in the chloroplast during electron transfer from photosystem II to photosystem I. However, the direction of H^+ movement in the mitochondria is towards the outerside 'he membrane while in chloroplasts it is towards the innerside.

Although, several hypotheses regarding ATP generation in mitochondria or chloroplast have been proposed, the most popular one is the *chemiosmotic hypothesis* proposed by Peter Mitchell in 1961. The evidence for this hypothesis has been provided by many others including A. Jagendorf and G. Hind and it is now the accepted theory of ATP generation. The fundamentals of the chemi-osmotic theory is the generation of a proton gradient across the inner mitochondria membrane, during electron transport from NADH to O_2. Six protons are generated per electron pair transported. These protons are secreted outside the inner mitochondrial membrane in the space between inner and outer mitochondrial membrane. The inner membranes have a specific orientation and they allow the secretion of protons only towards their outside and not towards inside. The proton gradient across the membrane generates sufficient electrochemical energy to drive oxidative phosphorylation, which is visualised to occur through a reversible reaction catalysed by ATPase;

$$ADP + Pi + H^+ \underset{ATPase}{\longleftrightarrow} ATP + H_2O + H^+$$

The enzyme can cause the hydrolysis of ATP to ADP and Pi as well, and the equilibrium constant favours the hydrolysis of ATP rather than its synthesis. However, Mitchell pointed out that if H_2O was removed from within the approach of catalytic site of ATPase then ATP will not be hydrolysed and according to law of mass action, synthesis of ATP will be favoured. According to his theory, H_2O is removed from the site by transport of OH^- to the one side of the membrane (innerside) and H^+ to the otherside (outer side). These ions on either side can combine with the counter ions (H^+ or OH^-) to form H_2O molecules. Several factors present in the membrane seem to be involved in creating this proton gradients. These factors or molecules are oriented in such a way that protons are secreted outside and OH^- inside. The energy for this section or transport and asymmetric distribution of protons comes from the electron transport system.

The proton gradient creates a proton motive force (pmf) which can be expressed in terms of free energy; that is, the energy which can do work. The free energy ultimately drives the synthesis of ATP from ADP and Pi.

The most important proof for chemi-osmotic theory comes from the use of chemicals in phosphorylation studies. The chemicals which destroy pH (or proton) gradient across mitochondrial or chloroplastic membranes prevent ATP synthesis. These chemicals are called uncouplers. Dinitrophenol is one of these. Further ATP is synthesized when a pH gradient is imposed on mitochondria or chloroplasts in the absence of electron transport.

Cyanide Resistant Respiration: As mentioned earlier in this chapter, cyanides and also the azides and carbon monooxide, inhibit respiration. It is believed that this inhibition is due to the effect on enzyme terminal oxidase, which catalyses the electron transfer between cytochrome a_3 and O_2. The effect ultimately leads to the inhibited O_2 uptake and ATP generation.

However, in some plant tissues, there is continued respiratioh in the presence of cyanides, azides or carbon monooxide. For example, aged potato slices show very little cyanide inhibition of respiration. Several algae, fungi and a few bacteria and animals are also resistant to these poisons. These organisms seem to have an alternate branch in the electron transport system, which is insensitive to cyanide, etc. This alternate branch allows the transport of electrons to O_2. Although, the details of alternate pathway are not known, it is assumed that the electrons from CoQ are transported to O_2 through an alternate oxidase enzyme. In this transfer no ATP is formed but only heat is evolved. The only ATP formed is prior to CoQ in complex I.

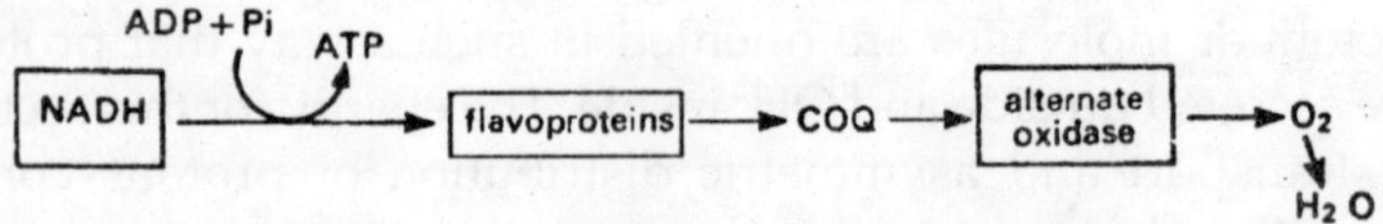

Electron transport system in cyanide resistant respiration.

Cyanide insensitive respiration is most apparent in the spadix of some lily flowers belonging to the family Araceae. The spadix in *Arum* and *Suaromatium* is a highly specialised structure. The rate of respiration in spadix is so high that temperature inside may rise by 10 to 15°C. This high heat causes the evaporation of odoriferous compounds such as amines, ammonia, indole, etc., when the spadix opens. These compounds attract insects for pollination. Thus, the high heat generating alternate respiratory pathway in these tissues, is of survival value to the plants. This alternate Pathway is stimulated by cyanides and other such poisons.

Respiratory Quotient

The respiratory quotient (R Q) is defined as the ratio of CO_2 evolved to O_2 taken in during respiration. When a hexose (glucose or fructose) molecule is completely oxidised involving the pathways of glycolysis and TCA cycle 6 molecules of CO_2 are produced and 6 molecules of O_2 are utilized in the process;

$$C_6H_{12}O_6 + 6O_2 \rightarrow 6CO_2 + 6H_2O$$

R Q in this case will be equal to: $\dfrac{\text{number of moles of } CO_2}{\text{number of moles of } O_2}$

$$\therefore \frac{6}{6} = 1$$

Most tissues which utilise carbohydrates as respiratory substrate have R Q close to this value *i.e.*, 1. For example, R Q measured for most green leaves averages about 1.05.

However, R Q value may vary considerably according to the tissues as they store and utilise various substrates in respiratory metabolism. A number of other biochemical reactions involving CO_2 and O_2 also effect the R Q value. When carbohydrate rich seeds germinate, the R Q is not exactly 1.0 but it is slightly higher. For example, in wheat seedlings it is around 1.3. This is because, organic acids are also utilised as respiratory substrates besides carbohydrates in these tissues.

The complete oxidation of citric acid for example, will show a R Q value of 1.33;

$$C_6H_8O_7 + 4.5\ O_2 \rightarrow 6CO_2 + 4H_2O$$

Citric acid

$$R\ Q = \frac{6}{4.5} = 1.33$$

Similarly when malic acid is oxidised R Q is 1.33 and when succinic acid is oxidised R Q is 1.14;

$$C_4H_6O_5 + 3O_2 \rightarrow 4CO_2 + 3H_2O$$

Malic acid

$$R\ Q = \frac{4}{3} = 1.33$$

$$C_4H_6O_4 + 7O_2 \rightarrow 8CO_2 + 6H_2O$$

Succinic acid

$$R\ Q = \frac{8}{7} = 1.14$$

Many seeds store fats or oils, which are used as respiratory substrate during germination and early growth of the seedling. In these cases, the value of R Q is often less than one. For example, if the oleic acid is the predominant fatty acid in the seed, R Q will be 0.71;

$$C_{18}H_{34}O_2 + 25.5\ O_2 \rightarrow 18CO_2 + 17H_2O$$

Oleic acid

$$R\ Q = \frac{18}{25.5} = 0.71$$

For palmitic acid, the R Q will be 0.68;

$$C_{16}H_{34}O_2 + 23.5\ O_2 \rightarrow 16CO_2 + 17H_2O$$

Palmitic Acid

$$R\ Q = \frac{16}{23.5} = 0.68$$

Actual measurements of R Q in germinating castor seeds, indicate that R Q decreases to as low as 0.4, or even less. This is because besides fatty acid oxidation, the germinating castor seeds have an efficient conversion of fatty acids to sucrose. The R Q for conversion of palmitic acid to sucrose is 0.36.

In some species, R Q varies according to day night conditions as is most apparent in succulents. These species fix CO_2 in dark through CAM to synthesize organic acids. During this period most of the respiratory CO_2 is used up in CAM and R Q value comes to less than one and very often zero. However, when organic acids are in abundant supply and CAM is restricted during the day time, the organic acids are oxidised and R Q is more than one.

Factors Affecting Respiration: A number of internal and external factors are known to influence the aerobic respiration of plant organs or tissues. The rate of respiration which is measured either by the rate of O_2 consumption or by CO_2 evolution, is also influenced by other metabolic processes and general physiological status of the plant. The effect may be either on the exchange of gases (O_2 and CO_2) reflecting the effect on stomatal opening and closing or on the metabolic process itself.

Plant Age and Tissue Types: Generally the younger tissues have higher rate of respiration than the older ones. The mature and senescent tissues have a very low rate of respiration. This is natural because these tissues have very low anabolic activities and therefore require less energy for other cellular activities.

However, when the rate of respiration of a single leaf is measured at different ages, it shows a typical sigmoidal curve. The rate of respiration is low in very young leaf and it increases in the initial stages of the development of leaf and then it reaches a maximum when leaf is fully mature. The rate of respiration declines after that.

In many mature leaves and fleshy fruits which also show a characteristic sigmoidal curve with respect to age, the rate of respiration increases temporarily just before the onset of senescence or ripening of the fruit. This rise in respiration have been called *climacteric.* The rise in respiration in fruits has been ascribed to the increased ethylene production, which is known to induce respiration and senescence. Examples of fruits which show a climacteric rise in respiration are apple, pear, tomato, banana, etc. Some examples of fruits which do not show climacteric rise are citrus fruits, pineapple and strawberry.

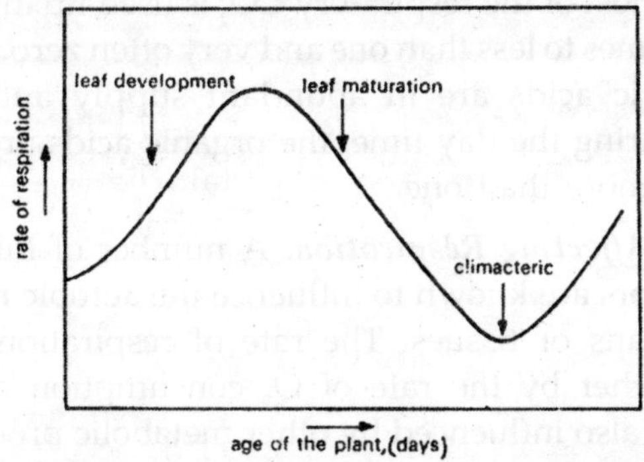

The rate of respiration depends upon the tissues as well. Respiration rate of inactive tissues such as older roots and leaves, resting fruits' or stems is lower than the actively growing or metabolically active tissues. Dormant seeds also have a very low rate of respiration, but the rate increases dramatically once they imbibe water and start germinating. Tissues like spadix of *Arum* lily have a very high rate of respiration. Typical rates of respiration of some plant tissues are given in table.

Rate of Respiration of Some Plant Tissues

Tissue	*Respiration rate, μ moles O_2 h^{-1} g^{-1} dry weight*
Dry seeds	0.005
Young seedlings	60-70
Beet slices	50
Tomato root tip	300
Arum spadix	2,000
Whole plant of a herb	3-5

Availability of the Substrate: As a general principle, the rate of respiration increases with an increase in substrate (specially the carbohydrates) availability to the respiring tissues. The etiolated leaves from dark grown seedlings have a low rate of respiration than the green actively photosynthesising leaves. However, if the etiolated leaves are floated on a carbohydrate (example sucrose) solution, the rate of respiration increases. Further in a canopy, the lower, shaded leaves have a low respiration rate as compared to the upper leaves exposed to bright light.

Temperature

Respiration like other biochemical processes is affected by temperature. In general, within certain limits, increase in temperature results in an increase in respiration (Fig.). The rate reaches a maximum value at optimum temperature and then it declines. Usually the value of Q_{10} between 0 and 25°C is between 2 and 3. That is to say an increase in temperature within this range causes an increase in respiration by 2 to 3 folds. At about 25°C there is often a decrease in Q_{10} and the value may reach about 1 at 30-35°C At above 35°C, there is a decline in respiration and the value of Q_{10} is less than 1. This decrease in respiration is apparently because of the inactivation and destruction of enzymes involved in respiration. This may also be due to limited availability of O_2 as its solubility and diffusion decreases at higher temperature.

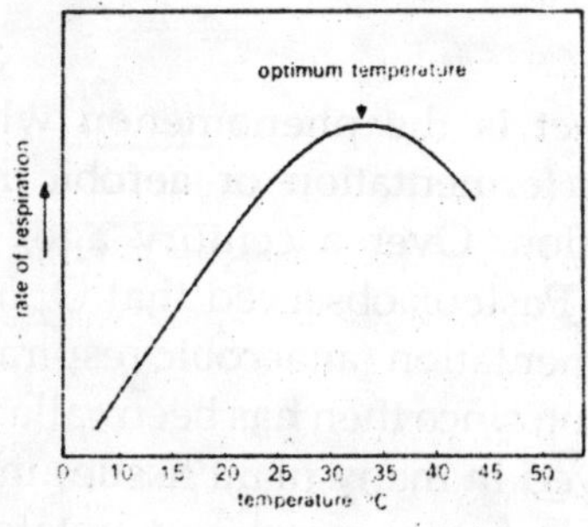

The effect of temperature on rate of respiration.

The optimum temperature for respiration varies according to the species and also according to the prevailing environment during the plant's growth. But in most tropical species, it is between 30 and 35°C. The xerophytes have usually higher optimum temperature than the mesophytes.

The rate of respiration is inhibited at very low temperature as well. In most species the rate of respiration is almost zero at freezing temperature. However, in some plant tissues the respiration has been recorded at as low as -20°C.

Oxygen Concentration

Oxygen is essential for aerobic respiration, as it is one of the reactants in the metabolic process. It is the ultimate acceptor of electrons and protons in electron transport system of the mitochondria. The effect of O_2 concentration of the atmosphere varies according to the tissues and prevailing conditions. However, the cytochromes involved in electron transport system have a very high affinity for O_2, and therefore they can function in oxidative metabolism even at very low concentration of the O_2. Variation in atmospheric O_2 concentration by a few percent (the atmospheric level is 21%) has no effect on the rate of respiration. In fact it has been observed that in most tissues the rate of respiration is uniform between 3 and 21%. Very high concentration of O_2 will inhibit respiration perhaps by oxidation of necessary respiratory components such as substrates or enzymic proteins.

Pasteur Effect

Pasteur effect is the phenomenon which relates to the effect of O_2 on fermentation or aerobic respiration and on aerobic respiration. Over a century ago, the French micro-biologist, Louis Pasteur observed that O_2 inhibited utilization of sugars in fermentation (anaerobic respiration) by yeast cells. This phenomenon since then has been called Pasteur effect and has been observed in many plant tissues including barley and pea leaves, apple fruits, carrot and potato tubers.

When the plant tissues are kept in complete anaerobic conditions, the process of fermentation or anaerobic respiration is the primary source of CO_2 evolution and energy generation. However, when the tissues are transferred to aerobic conditions, the fermentation or anaerobic respiration and any CO_2 or energy generation through this process is suppressed. At the same time, O_2 favours aerobic respiration, which increases with increase in O_2 concentration to some extent. A typical example of Pasteur effect, as observed in apple fruits is illustrated in figure.

As O_2 concentration is increased from zero level, the rate of respiration as measured by CO_2 evolution decreases drastically until about 1-2% of O_2 is reached. This is because anaerobic respiration is suppressed. The presence of O_2 interferes with the glycolytic process. A further increase in O_2 (above 1-2%) increases respiratory CO_2 evolution to certain extent. This is because increasing O_2 concentration increases aerobic respiration by stimulating the TCA cycle activity. However, any CO_2 evolution due to anaerobic respiration is insignificant at this stage.

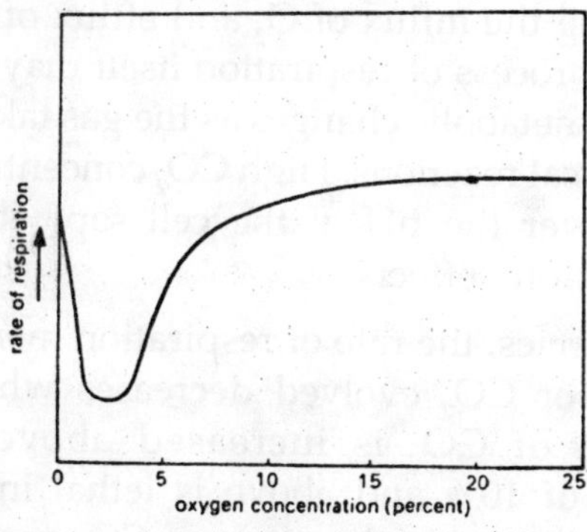

Pasteur effect in apple fruit respiration.

Several scientists have tried to investigate the molecular mechanism of Pasteur effect in yeast cells and in animal and plant tissues. The effect seems to be related to the allosteric control of the enzyme phosphofructokinase by ATP, ADP, Pi and citric acid. This enzyme is the rate limiting enzyme in glycolysis.

Among other things, the enzyme is inhibited by ATP and is stimulated by Pi þy allosteric modifications. Under anaerobic conditions, when there is more of pi and less of ATP, the enzyme is stimulated and glycolysis is increased. On the other hand, when there is sufficient build up of ATP, the enzyme is inhibited and glycolysis is decreased. However, there seem to be other points of control as well in glycolysis by ATP, ADP and Pi.

Pasteur effect, is an important factor in fruit storage where the objective is 10 decrease sugar loss due to respiratory activity during the storage. This is done carefully by decreasing the atmospheric O_2 to a level where anaerobic respiration is minimum, but where sugar breakdown is not stimulated by aerobic process.

Carbon Dioxide Concentration: The effect of CO_2 concentration on respiration may be either through the stomatal opening and closing or due to its effect on the process itself. High CO_2 concentration is known to close stomata while lower concentrations stimulate opening in most species. Stomatal opening and closing may regulate the gaseous exchange in plants including the influx of O_2 and efflux of CO_2. The effect of CO_2 on the process of respiration itself may be indirect and through other metabolic changes as the gas takes part in many other biochemical reactions. High CO_2 concentration inside the tissue may lower the pH of the cell sap which may cause general deleterious effects.

In most species, the rate of respiration measured by either O_2 consumed or CO_2 evolved decreases when atmospheric concentration of CO_2 is increased above about 5%. A concentration of 10% and above is lethal in most cases. In potato tubers, however, the effect of CO_2 varies according to the exposure period. Exposure to high CO_2 for shorter periods decreases respiration rate while high CO_2 for longer (24 h or more) periods, increases respiration.

The inhibition of respiration by high CO_2 in the natural environments is most apparent in roots and underground stems. The soil atmosphere is often rich in CO_2 because of microbial and edaphic bio-activities.

Tissue Hydration

Most tissue contain about 80-95% of water, except for the dry seeds. The respiration rate generally decreases when water content of the tissue decreases, although in a few cases there is a transient rise in respiration during mild water stress. However, minor variations in soil water content normally do not affect the hydration effects of the plants and have no effect on respiration. For example, in cereals such as barley, wheat or maize it has been found that the respiration rate does not decrease until soil water stress reaches to about -10 bar. It should be noted however, that many other Physiological processes such as leaf elongation in the cereals is inhibited even at -2 bar.

The Pollutants

Like photosynthesis, respiration is also inhibited by most of the environmental pollutants. Imbibition of respiration by heavy metals such as Pb and Cd is believed to be due to inactivation of the enzymes by the metals. The metals can bind with the sulfur containing amino acids of the enzymes. Air pollutants SO_2, NO_2, and O_3 are also known to inhibit the rate of respiration. However, very low concentration of SO_2 and NO_2 (less than 0.2 ppm) are known to increase respiration in some species such as soyabean and pintobean.

The increased respiration is believed to be due to increased cellular activities for metabolizing the pollutants to non-toxic forms. Respiration is vital to supply energy and metabolites not only for normal metabolism and growth, but also for repair and detoxification. The inhibitory effects of the air pollutants on respiration is believed to be due to general membrane damage and acidity. The acidity is known to influence the electron flow and phosphorylation.0

7

TRAINING BY PRACTICE

- *Leagans:* "Extension education is an applied science consisting of contents derived from researches, accumulated field experiences and relevant. Principles drawn from the behavioural sciences, synthesised with useful technology, in a body of philosophy, principles, contents and methods focussed on the problems of out of school education for adults and youths".
- "Agricultural extension is a bridge that fills the gap between agricultural research stations on the one hand and the farming population on the other by the establishing a suitable teaching organisation at various levels of administration."
- "Extension education is a behavioural science following a continuous, persuasive and discriminating educational process, it aims at affecting the behavioural components of people in a desirable direction, through conviction, communication and diffusion by its Proven methods, principles and philosophies resulting in learning involvement of both client and change-agent systems."

TRAINING PRINCIPLES

The principles discussed below one by one

(1) Principles of interest and needs.

(2) Grass-roots principle of organisation.

(3) Principle of cultural differences.

(4) Principle of cultural change.

(5) Principle of Cooperation and participation.

(6) Principle of applied science and democratic approach.

(7) Principle of learning by doing.

(8) Adaptability principle in the use of extension teaching methods.

(9) Principle of trained specialists.

(10) Principle of leadership.

(11) Whole family principle.

(12) Principles of satisfaction.

Extension by Diverse Agencies and Organisation

(a) Work at Shantiniketan: In 1908, Shri Rabindra Nath Tagore, under his scheme of rural development work, started youth organisations in the villages in the Kaligram pargana of his zamindari.

- In 1921, he established a rural reconstruction institute at Shantiniketan.

Objectives: (1) To create a real interest in people for rural improvement. (2) To study rural problems. (3) To improve village sanitation.

(b) Gurgaon Experiment: Rural welfare movement on a mass scale was first started by Mr. F L. Brayne, Deputy commissioner in the Gurgaon district of the Punjab, adjacent to Delhi.

- Under his programme, a village guide was posted in each village who was to act as a channel through which the advice of experts in various departments could be passed on to the villagers.

(c) Marthandam Programme: Marthandam programme was set up under the guidance of Y.M.C.A. (young men's christian association) in Travancore.

- The leader in this work was Dr. Spencer Hatch, an American agricultural expert. The essential technique of the centre was 'self help with intimate expert counsel.'
- From the demonstration centre at Marthandam, about hundred villages were covered through Y.M.C.A centres in villages.

(d) Etawah Pilot Project: The idea of starting this project was conceived and born in 1947, it was put into action with headquarters at Mahewa village about 17 miles form Etawah (U.P.) in September 1948.

- Mr. Albert Mayer of U.S.A. who came to India with the American forces in 1944 was the originator of the project.

He started it with the aim of introducing intensive work on the rural reconstruction front.

(e) Indian Village Service (I.V.S.): Its founder was Mr. Arther T. Mosher of New York and Shri B.N. Gupta who established it in 1945.

- The objectives of this organisation were to assist village people to realise the best in their own villages by developing individuals, volunteer leaders and local agencies and enabling them to be effective in helping themselves and others.

(f) Sarvodaya Programme: It was a Gandhian concept and evoked great enthusiasm in Bombay state.

- The main features were simplicity non-violence, holiness of labour and reconstruction of human values.
- Its aimed at raising the standard of living, scientific development of agriculture, promotion of cottage

industries, spread of literacy, medical and health facilities and the development of village panchayat.

(g) Firka Development Scheme of Madras State: It was Government sponsored and aimed adoptment of the Gandhian ideals of Gram Swaraj by bringing about not only educational, economic, sanitary and other improvements in villages but also by making the people self-confident.

- The scheme was launched in the last quarter of 1946 in 34 Firkas throughout the state and on April 1, 1950, it was extended to another 50 additional Firkas.
- The collector, who was primarily responsible for the successful working of the scheme in the district.

(h) Nilokheri Experiment: It was originally started to set up again 7000 displaced persons from Pakistan and later integrated with the 100 surrounding villages into what came to be a rural cum urban township.

- It was built round the vocational training centre that was transferred from Kurukshetra in July, 1948 to the 1100 acres of swampy land on the Delhi-Ambala highway.

(i) Intensive Agricultural Development Programme (I.A.D.P): The intensive agricultural development programme was popularly known as a package programme. This name was given because of the collective and simultaneous application of all improved practices.

- These practices were improved seeds irrigation, fertilizers, plant protection, implements, credit etc.
- The programme was started in July 1960 in 7 selected districts in various states.

These were:

(1) West Godavari (A.P.) (2) Shahabad (Bihar)
(3) Tanjore (Madras) (4) Raipur (M.P.)

(5) Ludhiana (Punjab) (6) Pali (Rajasthan)

(7) Aligarh (U.P.)

Developmental Process

Extension education centres around three units

(a) Farmers or rural peoples

(b) Innovations or inventions

(c) Extension workers

- The technological development of education can be related to various fields of extension like Agriculture and Home Science.

Concept of Agricultural Extension

Agricultural extension is concerned with agricultural education assisting farmers to bring about continuous improvement in their physical, economic and social well being through their individual and Cooperative efforts.

It makes available to the farming community the scientific and other factual information the training and guidance for the solutions of the problems in agriculture including Animal Husbandry, Gardening, Agricultural Engineering etc.

- Role of Agricultural Extension: Each country has to develop its extension organisation according to its own social cultural, economic situations at a given stage in its development. There are however, certain broad generalisation which can made it. These are

  (1) Extension work can be most effective only when it is entirely educational in nature and free from regulatory and other govt. responsibilities.

  (2) As development proceeds, more specialisation at the field level may be feasible and desirable.

  (3) The successful work of the extension officer at the field level depends upon their being supported by specialists at the next higher level of administration.

(4) Efforts should be made to keep extension and research closely related to each other.

(5) Extension work must be directly towards the family as a whole men, women and children.

- Work with youth and farm women is important.

(6) In the interest of providing maximum effective service within the limited resources, there is an urgent need for effective coordination of all the agencies serving rural people. (IFS 2001)

Communication Perceptions

Leagans: "It is a process by which two or more people exchange ideas, facts, feelings or impression in ways that each gain a common understanding of the message in essence it is the act of getting a sender and a receiver tuned together for a particular message or series of message"

"Communication is the process of effecting an interchange of understanding between two or more people".

- Teaching and learning is the process of communication.
- Teacher acts as catalyst.
- Communication is a two way or double way process.
- Communication involves
  (a) Source and his message
  (b) Receiver and his response.
- The chief emphasis in communication on target audience.
- Communication systems

K—Knowledge

G—Generating

S—System

D—Disseminating

C—Consuming

Agencies

Universities
Research institutes
Private organisation
Voluntary organistion } KGS

- ***Mass Media***

Universities
Research Institutes
Govt. agencies
Voluntary organisations } KIDS

- ***Farmers***

In agriculture } KCS

Key Elements of Communication

(a) Sender

(b) Message

(c) Message treatment *i.e,* encoding

(d) Communication medium

(e) Receiver: It is the most important element because every element is directed towards receiver.

(f) Receiver's response.

Communication Models

Sender—Message—Treatment —Receiver—Channel

- Leagan's Model

Communicator—Message—Channel—Response of —Audience —Treatment — Audience

- According to Berlo

Source—Encoder—Message—Receiver—Decoder—Channel

Ramification of Communication

Communication media and extension methods are the tools of approach as they increase the effectiveness of extension work.

- An extension worker, must first know *"what methods are available to* him" secondly, he should know-"When *to* use a given *method* and thirdly *"he should become efficient in using each method"* The approaches are given below
- Approaches in communication and extension

(A) Individual approach through:

(1) Personal visits

(a) Farm visits (b) Home visits

(2) Personal letters

(B) Group approach through

(1) Demonstration

(a) Method (b) Result

(2) Training the leaders

(3) Discussion meetings

(a) Group discussion (b) Panel discussion

(c) Symposium (d) Forum

(e) Group interview (f) Workshop

(g) Dialogue or public conversation

(4) Field days

(5) Tours

(C) Mass approach through

(1) Films and slides: (a) Farm visits, (b) Home visits

(2) Flash-cards (3) Printed materials

(4) Radio (5) Recorded talks

(6) Charts, diagrams etc. (8) Circular letters

(9) Newspaper stories (10) Compaigns

(11) Television

Division According to Equipments

(1) Projected Aids: Here projection is controlled by motor and electricity *e.g.*

(a) Film Projector

(b) Slide Projector

(c) Overhead Projector: It is indirect projection and transparencies used.

(d) Opaque Projector: There is no need of transparency.

(2) Non-projected Aids: Here no machinery is required *e.g.* Flash cards, groups, charts, leaflets, folders, black boards.

Division According to Audio-visual Aids

(1) Audio: Radio, taperecorder

(2) Visual: Flash cards, overhead projector, slides etc.

(3) Audio-visuals: Puppet, drama, television, cinema etc.

Approach of Demonstration

Among the communication and educational techniques 'demonstration' are the oldest, best and simplest tools for transmitting sophisticated technology in a simple and understandable form.

- Demonstration means showing by doing. The basic principles of demonstration is learning by seeing and doing. There are three types of demonstration

Method Demonstration: It is the short-type demonstration. The method demonstration evolved by Dr. Seeman. A. Knapp. So the father of method demonstration is Dr. Seeman. A. Knapp. This is a single practice demonstration i.e., how to apply fertilizer in the field. This cannot compare between old and new technique or skill.

There are some points which should be taken into consideration while demonstrating practice.

- Phase of demonstration: An effective demonstration programme involves 5 essential phases
  (1) The first phase is of analysing the situation.
  (2) The second phase is deciding on goals to be accomplished in conducting a demonstration.
  (3) The third phase is the demonstration plan, and execution of demonstration.
  The plan includes the three stages:
  (i) Pre-demonstration stage
  (ii) Execution stage
  (iii) Follow-up or post-demonstration stage.
  (4) The fourth phase is evaluation of the demonstration programme.
  (5) The fifth phase is reconsidering after evaluation has taken place.
- Important tips conducting a good demonstration
  (1) The demonstrator should encourage the audience and see that people understand and learn to do what is being demonstrated.
  (2) Before starting the actual working, he should tell the audience.
  (3) Tell the audience especially the interested farmers.
  (4) Help the audience to go through the steps of demonstration so that they can repeat the process themselves without further help.
  (5) A demonstrator should be well-prepared in the subject himself and should take the help of his colleagues and teachers, whenever necessary.
  (6) Invite questions and create an atmosphere that the audience may clarify it doubts.

Result Demonstration: In this demonstration, the result of demonstrated practices are shown in comparison to existing practices.

- In result demonstration process, the extension worker may take the help of audio-visual aid like film, slides, pictures and comparative diagrams etc. So the farmers imagination is fired and they have some idea of how things were done to obtain higher yields.

Significant Points for Result Demonstration

(1) The students should demonstrate an specific problems of the area-drilling wheat against broadcast, hybrid maize against local variety.

(2) It should be carried out correctly, in detail and must be in sequence.

(3) Better to conduct more than one demonstration in the same village, on the same subject.

(4) Never try to discover new truths.

(5) Do not repeat demonstration repeatedly.

(6) Use audio-visual aids to support the result of demonstration.

(7) The demonstration should be located by a well-travelled roadside or path.

Visual Aids: Visual method of communication is more effective than other. Out of the five physical senses through which we learn, the eye is the most effective (helpful) in learning.

Planning for use of Visuals: For the use of our visuals we should think of: objective, audience, media, resource and technique.

- The subject matter of a visual is picked up after establishing the objectives of the visual.

Visual Aids-non Projected: The non-projected visual aids are divided into two groups: (A) Teaching aids (B) Display visuals.

(A) Teaching Aids: Teaching aids are non projected visual aid and can be prepared and used by teachers or extension workers are

(1) Chalk board

(2) Pictures, including photographs and hand drawn pictures.

(3) Flash cards

(4) Flip books - similar to flash cards

(5) Lip charts -few charts on a particular theme put together and bound together at the top.

(6) Flannel graph.

(B) Display Visuals: Displays visuals is not projected aid. It is also non-projected visual aid. Its examples given below

(1) Posters: A poster helps the extension workers to get across one idea to the audience. It is a good communication method.

(2) Diagrams: We can draw a picture of any speciman like goat, cow, building plan etc., it is called diagram.

(3) Charts: A chart contains a series of idea. Generally poster is used in the awareness stage but a chart is used in the testing stage. Chart can be

(a) Pictorial chart (b) Organisational chart
(c) Flow chart (d) Suspense chart
(e) Stream chart

(4) Graphs: For getting a precise idea or drawing a comparison or statistical data needs presentation of idea through graphs. Different types of graphs found there

(i) Line graphs (ii) Tree or stream graphs
(iii) Area graphs (iv) Bar graphs
(v) Pie graphs (vi) Photographs

(5) Flip Chart: It is like an album of drawings, pictures, charts. It is an assists an extension workers or teachers to tell story.

(6) Flash Cards: These are the series of cards which when presented before the audience in proper sequence, tell a complete story.

Each card is of about 10" to 12" in size and contain a picture or diagram.

(7) Felt Board: This is a board with a surface of felt, flannel or khaddar cloth which is used for fixing pictures or parts of material on the back of which there is piece of sand paper to stick

(a) Chalk Board: Even it is called blackboard. This is helpful in meeting and group discussions. This is generally used in villages and other places.

(b) Bulletin Board: It is a board with a background of coloured cloth. It can be covered with glass, netting or a plane board. The material for display may be news sheets, booklets, cartoons, circular letters.

- This is a good medium for displaying the creative work of the students.

(c) Exhibitions: Exhibition is one of the best programme of extension which the extension workers can arrange successfully. An exhibition can show actual things, practices, programmes in progress through charts, diagrams, displays, models etc.

Visual Aids Projected

Such aids include slides, film-strips and films. The projection of these aids involves various types of equipment beginning from the simple magic lantern to the complicated sound film projectors.

Types of projection: (1) Direct Projection, (2) Indirect Projection, (3) Reflected Projection Projectors

(1) Opaque Projectors: In India and England, opaque projector is also called as Epidiascope.

(2) Slide Projectors: Slide projectors are based on principle of direct projection.

(3) Overhead Projector: This equipment uses the principle of indirect projection.

(4) Movie Projectors: Movie projector may be classified on the basis of size of film.

Audio-aids Sound Recording

There are three methods of recording the sound:

(1) Mechanical

(2) Magnetic process

(3) Optical process

Disc recording is done by the mechanical process.

- Tape and wire recording is done by the magnetic process.
- Recording of sound on a movie film by the optical process.

Disc Recording: In the discs or gramophone records, discs of the lac and plastic are usually used. The rate of revolution per minute of the disc are 45, 78 or $33\frac{1}{2}$. The $33\frac{1}{2}$ r.p.m. records are long playing having 16 inches diameter and taking 15 minutes for each side.

Wire Recording: The wire employed for this purpose is a steel wire which has diameter of 4-5 mils. The process of magnetic recording on this wire is based on the working principle of an electro-magnet.

- This wire is available in spools which are of 3 types:

(1) lasting for 15 minutes

(2) $\frac{1}{2}$ hour

(3) one hour

Tape Recording: Using the principle of magnetic process recording can be done on tapes made of paper, plastic or metal.

The tape recorder accommodate two spools, one empty take-up spool and other containing tape for recording.

Radio: Radio is the good source of communication of ideas to the rural people especially in a country, like India. It is the good source of extension of agricultural information to farmers. The All India Radio stations publish a Journal Akashvani which gives the metres and frequency in kilo cycles on medium and short waves.

Audio-visuals, Non-projected

Puppets: Puppets can be very interesting aids for telling stories to the village people the puppets may be:

(1) Hand puppets, (2) String puppets,

(3) Rod puppets, (4) Shadow puppets

Drama as Extension Education

This is a good source of entertainment and extension education for the villagers. The students of the extension wings of agricultural with the help of their teachers, teachers of village schools, VLW's and villagers can arrange to produce dramas in their villages.

Talks: Even talk a good source of communication for this purpose should be good speaker. Confidence and courage are necessary for making a good speech.

Meetings: For this process of communication three stages given below :

(A) Planning

(1) Selection of topic
(2) Timing
(3) Selection of place
(4) Selection of speaker and chairman
(5) Advertising and publicity
(6) Physical arrangements

(B) Conducting the Meeting

(1) Programme procedure (2) Audience participation

(C) Follow-up of Meeting

(1) Summaries (2) Press reports
(3) Displays (4) Other method
(5) Evaluation

Kinds of Meeting

(1) Lecture forum (2) Panel discussion
(3) Symposium (4) Group interview
(5) Public conversation (6) Debate discussion
(7) Workshop

Participation in Conferences

We can participate in conferences more effectively by keeping the following tips into consideration

(1) Tell the audience in advance how many points you intend to make and specify when you come to each of these points.

(2) Be confident that you can do a good job.

(3) Speak naturally in a loud, clear but not a shouting voice.

(4) If possible, put an outline of points on a blackboard or chart.

(5) If the situation permits, speak to some of the audience members personally.

(6) Use only brief references and notes if possible avoid reading it.

(7) Take every possible opportunity to speak.

(8) Avoid giving a talk right after a meal. Mental capacity is at lowest after eating.

(9) Give an eye-catching title to your talk.

Significance of Tours

A tour is essentially a visit to demonstrations, plots, farms of progressive farmers, dairy, model house etc.

Objective of Tour

(1) To see the results of a new practice.

(2) To see a new practice demonstrated.

(3) To see the operation of a new tools.

(4) To see the accomplishment of other village, college, farms etc.

Communication through Written Words: It is playing a great role in communication with farm people, home, markets.

- Following methods for written communication:

  (1) Newspaper (2) Stories

  (3) Magazine articles (4) Leaflets

  (5) Pamphlets (6) Circular letters

  (7) Wall news papers
- Leaflets contains 4-12 pages stiched or stapled at the centre.
- Pamphlet varies in size form 12 to 24 pages.
- A bulletin is a bulkier publication with 24-48 or more pages.
- Circular letter is a quick and cheap method of communication and proves very effective if done well.
- Circular letters can teach villagers easily.

BIBLIOGRAPHY

Adam, N. : *Encyclopaedia of Plant Physiology,* Springer Verlag, Berlin, 1975.

Alam, B. : *Advanced Plant Physiology,* Longman, London, 1984.

Anderson, Edgar : *Plants, Man and Life,* University of California Press, California, 1967.

Arvill, R. : *Man and Environment : Crisis and the Strategy of Choice,* Penguin, New York, 1967.

Bailey, L. H. : *Hand Book of the Multiplication of Plants,* Discovery Publishing House, New Delhi, 1999.

Beevers, L. : *Nitrogen Metabolism in Plants,* Edward Arnold, London, 1976.

Bhatia, B. M. : *Poverty, Agriculture and Economic Growth,* Vikas Publishing House, New Delhi, 1977.

Billihgs, W. D. : *Plant, Man and the Ecosystem,* MacMillan, London, 1971.

Bowling, D. J. F. : *Uptake of Ions by Plant Roots,* Chapmann and Hall, London, 1976.

Bray, C. : *Nitrogen Metabolism in Plants,* Longman, London, 1984.

Candolle, Alphonse De : *Origin of Cultivated Plants,* Hafner, New York, 1886.

Crafts, A. S. and C. E. Crisp, : *Phloem Transport in Plants,* W.H. Freeman, San Francisco, 1973.

Dadiani, N. K. : *The Survey of Indian Agriculture*, Kasturi & Sons Ltd., Chennai, 1999

Deckock, P. C. : *Mineral Nutrition of Plants : a Symposium*, University of Oxford, New York, 2002.

Dennis, D. T. and Turpin, D. H. : *Plant Physiology, Biochemistry and Molecular Biology*, Longman, London, 1990.

Douglas, J. S. : *Advanced Guide to Hydroponics*, Pelham Books, London, 1976.

Dubey, R. S. : *Agricultural : Issues and Application*, Gyan Books, New Delhi, 1987.

Dunn, Edgar S. : *Location of Agricultural Production*, University of Florida Press, New York, 1954.

Evans, L. T. : *Day Length and the Flowering of Plants*, W.A. Benjamin, California, 1975.

Farmer, J. B. : *Study of Plant Life*, Discovery Publishing House, New Delhi, 1993.

Frankel, F. R. : *India's Green Revolution : Agriculture Costs of Land Growth*, Princeton University Press, Princeton, 1971.

Ftzler M. E. : *Plant Lectins : Molecular and Biological Aspects*, Academic Press, New York, 1985.

Gangawane, L. V. : *Vistas in Mycology and Plant Pathology*, Commonwealth Publishers, New Delhi, 2000.

Gauch, H. G. : *Inorganic Plant Nutrition*, Hutchinson and Ross, London, 1972.

Green, J. R. : *An Introduction to Vegetable Physiology*, MacMillan, London, 1911.

Gregor, Howard F. : *Geography of Agriculture : Themes in Research*, Prentice Hall, New Jersey, 1970.

Gupta, G.P. : *Test Book of Plant Diseases*, Discovery Publishing House, New Delhi, 2001.

Hansen, E. : *Post Harvest Physiology of Fruits*, Praeger Publishers, New York, 1966.

Hewitt, E. J. and Cutting, C. V. : *Nitrogen Assimilation of Plants*, Academic Press, London, 1982.

Hussain, M. : *Systematic Agricultural Geography*, Rawat Publications, Jaipur, 1999.

Isser, G. P. : *Changing Face of Poverty and Globalisation*, Gyan Books, New Delhi, 2002.

James, A.: *Weather and Agriculture*, Pergamon, Oxford, 1967.

Joshi, H. L. : *Agricultural of India*, Rawat Publications, Jaipur, 1990.

Karanth, Apoorva : *Plant Genetics and Food Crop Science*, Dominant Books, New Delhi, 1999.

Kirkby, E. A. : *Nitrogen Nutrition of Plants*, University of Leeds, Leeds, 1970.

Kramer, A. L. : *Principles of Biochemistry*, Worth Publishers, New York, 1982

Kramer, P. J. : *Plant and Soil Water Relationship : A Modern Synthesis*, McGraw Hill Book Company, New York, 1969.

Kumar, Arun : *Agricultural Development : Today and Tomorrow*, Anmol Publications, New Delhi, 1995.

Kumar, T. : *History of Rice in India : Mythology, Culture and Agricultural*, Gyan Books, New Delhi, 1988.

Laxmi Narasaiah, M. : *Agricultural Production*, Discovery Publishing House, New Delhi, 1999.

Mazliak, P. : *Lipid Metabolism in Plants*, Chapmann and Hall, London, 1973.

McKee, H. S. : *Nitrogen Metabolism in Plants*, Oxford University Press, London, 1962.

McNeil, M. Darvill : *Structure and Function of the Primary Cell Walls of Plants*, McGraw, London, 1984.

McRae, S. C. and Burnham, C. P. : *Land Evaluation*, Oxford University Press, New York, 1981.

Meidner, H. and Sheriff, D. W. : *Water and Plants,* Blackie and Sons, Glasgow, 1976.

Mengel, K. and Kirkby, E.A. : *Principles of Plant Nutrition,* Praeger Publishers, New York, 1974.

Mishra, S. R. : *Bacterial Plant Diseases,* Discovery Publishing House, New Delhi, 2003.

Misra, B., Kar, G. C. and Misra, S. N. : *Agro-Industries and Economic Development,* Deep & Deep Publications, New Delhi, 2002.

Mitter, V. : *Growth of Urban Informal Sector,* Deep & Deep Publications, New Delhi, 1988.

Mukherji, Shekhar : *Land-Use Change, Environment and Sustainable Development,* Banaras Hindu University, Varanasi, 2000.

Nomura, M. : *The Control of Ribosome Synthesis,* Edward Arnold, London, 1986.

Novikoff, A. B. : *The Endoplasmic Reticulum : a Cytochemist's View,* Oxford, New York, 1976.

Numa, S. : *Fatty Acid Metabolism and its Regulation,* Janis, London, 1984.

Nutman, P. S. : *Symbiotic Nitrogen Fixation in Plants,* Cambridge University Press, Cambridge, 1976.

Ogawa, Y. and R. W. King : *Plant and Cell Physiology,* Springer-Verlag, Berlin, 1990.

Opik, H. : *The Respiration of Higher Plants,* Edward Arnold Publishers, London, 1980.

Palladin, V. I. : *Plant Physiology,* Arihant Publishers, Jaipur, 1988.

Pannikkar, K. M. : *India and the Indian Ocean : An Essay on the Influence of Sea Power on Indian History,* George Allen and Unwin, London, 1951.

Ramshaw, J. A. M. : *Structure of Plant Proteins,* Oxford University Press, New York, 1982.

Robert, W. : *Middle America : Its Land and Peoples,* Prentice Hall, Englewood Cliff, 1989.

Roberts, J. A. and Tucker, G. A. : *Ethylene and Plant Development,* Butterworth, London, 2001.

Rubenstein, J. M., and Bacon, R. S. : *The Cultural Landscape : An Introduction to Human Geography,* Prentice Hall, New Delhi, 1990.

Salisbury, F. B. and C. W. Ross, : *Plant Physiology,* Woodworth Publishing Co., California, 1967.

Samanta, R. K. : *Agricultural Estension in Changing World Perspective,* Uppal Publishing House, New Delhi, 1995.

Shandhya, T. K. : *Agricultural Credit and Nabard,* Deep & Deep Publications, New Delhi, 2003.

Sharma, D. N. : *Minerals of India,* National Book Trust, New Delhi, 1994.

Sharma, S. S. : *Sustainable Agriculture : Poverty and Food Security,* Rawat Publications, Jaipur, 2002.

Sharma, V. K. : *Trees and Protection of Environment,* Deep & Deep Publications, New Delhi, 2004.

Siegelman, H. W. and G. Hind : *Photosynthetic Carbon Assimilation,* Plenum, New York, 1978.

Singh, D. K. : *Technological Change and Agricultural Development,* Deep & Deep Publications, New Delhi, 1993.

Singh, Harjinder : *Agricultural Problems in India,* Gyan Books, New Delhi, 1988.

Singh, L. : *Green Revolution and Cropping Pattern,* Deep & Deep Publications, New Delhi, 1993.

Singh, Surendra : *Agricultural Development in India : A Regional Analysis,* Kaushal Publications, Shillong, 1994.

Slatyer, R. O. : *Plant Water Relationship,* Academic Press, New York, 1967.

Somani, K. K. : *Indian Economics and Social Traditions,* Gyan Books, New Delhi, 2002.

Srivastava O. S. : *Agricultural Economics,* Rawat Publications, Jaipur, New Delhi, 1996.

Steward, F. C. : *Plant Physiology, A Treatise,* Academic Press, New York, 1963.

Street, H. E. and H. Opik : *The Physiology of Flowering Plants,* Edward Arnold, London, 1970.

Subrahmanya, S. : *Studies in Agricultural Development,* Deep & Deep Publications, New Delhi, 1996.

Sudhir, M. : *Applied Biotechnology and Plant Genetics,* Dominant Books, New Delhi, 2000.

Thomas, C. : *A History of the Land and Agriculture,* Prentice Hall, Englewood Cliff, 1978.

Valcarcel, M. : *Weather and Agriculture,* Oxford University Press, London, 1997.

Varner, J. E., Flint D. and R. Mitra, : *Characterization of Protein Metabolism in Cereal Grains,* Academic Press, New York, 1976.

Venkateshwarlu : *Developing Agricultural Technology,* Rawat Publications, Jaipur, 1998.

Verma, M. : *The Survey of Indian Agriculture,* Kasturi & Sons, New Delhi, 1999.

Verma, P. C. : *Surplus Manpower in Agriculture and Employment Policy,* Deep & Deep Publications, New Delhi, 1991.

Vyas, V. S. : *Policies for Agricultural Development,* Rawat Publications, Jaipur, 1997.

Wilkins, M. B. : *The Physiology of Plant Growth and Development,* McGraw Hill book, London, 1969.

Zeevart, J. A. D. : *Physiology of Flower Formation,* McGraw Hill, London, 1976.

Zimmermann, M. H. : *Encyclopeadia of Plant Physiology,* Academic Press, New Work, 1982.

INDEX

D

E

F

G

H

I

L

M

N

O

P

T

U

V

W

X

Z

□□□